AF595131

Ecological and Health Risk of Soils, Sediments, and Water Contamination

Ecological and Health Risk of Soils, Sediments, and Water Contamination

Editor

Zeng-Yei Hseu

MDPI • Basel • Beijing • Wuhan • Barcelona • Belgrade • Manchester • Tokyo • Cluj • Tianjin

Editor
Zeng-Yei Hseu
National Taiwan University
Taiwan

Editorial Office
MDPI
St. Alban-Anlage 66
4052 Basel, Switzerland

This is a reprint of articles from the Special Issue published online in the open access journal *Water* (ISSN 2073-4441) (available at: https://www.mdpi.com/journal/water/special_issues/soils_contamination).

For citation purposes, cite each article independently as indicated on the article page online and as indicated below:

LastName, A.A.; LastName, B.B.; LastName, C.C. Article Title. *Journal Name* **Year**, *Volume Number*, Page Range.

ISBN 978-3-0365-0034-8 (Hbk)
ISBN 978-3-0365-0035-5 (PDF)

Cover image courtesy of Zeng-Yei Hseu.

Contents

About the Editor

Zeng-Yei Hseu (Professor): Hseu is Professor of soil quality and environmental sciences at National Taiwan University (NTU), Taipei, Taiwan. He had been employed as a Guest Professor at Kyoto University in 2010 and at Meiji University in 2011, Japan. He had also been a Visiting Scholar at Hong Kong Polytechnic University in 2018. Professor Hseu served as the president of the Chinese Society of Soil and Fertilizer Sciences (Taiwan) in 2016–2019 and President of East and Southeast Asian Federation of Soil Science Societies (ESAFS) in 2018–2019. His major topics of interest are heavy metal dynamics and mineralogy of serpentine soil, morphology and genesis of wetland soil, soil chronosequences on river and marine terraces, and soil heavy metal contamination and remediation. Professor Hseu is the author or coauthor of approximately 100 scientific papers and book chapters.

Editorial

Ecological and Health Risk of Soils, Sediments, and Water Contamination

Zeng-Yei Hseu

Department of Agricultural Chemistry, National Taiwan University, Taipei 10617, Taiwan; zyhseu@ntu.edu.tw; Tel.: +886-2-33664807

Received: 10 October 2020; Accepted: 13 October 2020; Published: 15 October 2020

Abstract: Soils, sediments, and water require careful stewardship for the planet's security to achieve the Sustainable Development Goals (SGDs) set from the United Nations. However, the contamination of these natural resources can damage ecological and human health, and thus we need a comprehensive approach to provide a remediation reference for the SDGs. The aim of this Special Issue (SI) was to gather the papers emphasizing different aspects and findings of the contamination processes, remediation techniques, and risk assessment of soils, sediments, and water. The Guest-Editor of this SI collected seven papers dealing with biochar application for the reduction in soil nutrient leaching by Kuo et al. and for the immobilization of soil cadmium by Chen et al. Their works contributed to not only sustain soil functions but also to prevent sediments and water from contamination. Moreover, in situ stabilization by environmentally compatible approach is a green remediation of sediments such as thin-layer capping for freshwater and estuary sediments by Ou et al. and Ch'ng et al., respectively. Bioassays including microbiological response and enzyme activities were used to test water quality by Martín et al. and Aljahdali et al., in addition to the finding of antibiotic-degrading bacterial strains reported by Yang et al. in sewage sludge. These papers may aid to update and incorporate new views and discussion for the SDGs.

Keywords: bioaccessibility; biochar; biomarkers; green and sustainable remediation; heavy metal; SDGs; thin-layer capping

---

1. Introduction

There has long been concern about the issue of soils, sediments, and water pollution by various contaminants worldwide. Soil provides an interface between the lithosphere, atmosphere, hydrosphere, and biosphere, and thus improvement of soil function has recently become a major priority in ecosystems, particularly because of the growing awareness regarding the role of soil in controlling sediment and water quality crucial for human benefit [1]. For instance, the sustainable monitoring and management of contamination and remediation of soils, sediments, and water toward reaching the 17 Sustainable Development Goals (SGDs) set from the United Nations have been recognized as important in previous studies [2], which identified several targets with direct synergies with these natural resources across the goals.

Regarding soil and sediment remediation, conventional practices such as washing, landfilling, and excavation are commonly poor-feasible especially on a large scale because they are not environmentally compatible and are economically-prohibitive [3]. These concerns have prompted green and sustainable remediation (GSR) for the contamination of soils and sediments. Among GSRs, the in situ stabilization of contaminants using reactive or immobilizing materials has received increasing attention [4]. The aim of adding amendment is to sequester and stabilize contaminants in soils or sediments to reduce their ability to spread into water or biota, and thus to reduce their risk to human health. Aquatic ecosystems including sediments and water often play as the sinks of contaminants transported from soil contamination and wastewater discharge. To identify the impact of contaminants in water by bioassays, it is necessary to test different representatives of biomarkers as indicators of substances that are harmful to living cells and tissues, useful even in the cases where physicochemical parameters fulfill the requirements of water quality. This identification approach may coincide with the GSR principles of soil and sediment contamination for ecological and human health.

2. Overview of This Special Issue

Seven original papers are published in this Special Issue: two are the topics of soil remediation by using biochar, two are heavy metal stabilization by iron sulfide-based amendments in sediments, two are evaluation of the biomarkers of heavy metal contamination in river water, and one is biodegradation of antibiotics by specific bacterial strains screened from sewage sludge.

Biochar acts as a liming amendment in soils, increasing the retention capacity of nutrient and heavy metal in the soil solids. Thus, the application of biochar has received growing interest as a sustainable technology in contaminated soils because it boosts the intrinsic sorption capacity of the soil [5]. Kuo et al. evaluated the effects of biochar on organic carbon (OC) and nutrient retention and leaching in a coarse-textured soil [6]. They conducted a 42-day column leaching experiment by the tested soil mixed with 2% of biochar pyrolyzed from the wood sawdust of Honduran mahogany (*Swietenia macrophylla*) at 300 °C (WB300) and 600 °C (WB600). The results indicated that biochar application increased the final soil pH and OC, concentrations of ammonium-N, nitrate-N, and available phosphorus (P) but not exchangeable potassium (K) concentrations. They concluded the ability to retain N, P, and K in the tested soil differed with pyrolysis temperatures of biochar, but WB300 and WB600 effectively contributed to the conservation of groundwater and river water in the catchment.

Biochar from rice husk was applied into a cadmium (Cd) contaminated soil by Chen et al. [7]. Lettuce (*Lactuca sative*) and pak-choi (*Brassica chinensis*) were planted in the biochar-amended soil to observe the accumulation, translocation, and chemical forms of Cd in the leafy vegetables. In addition, the vegetable-induced hazard quotient was calculated via the chemical form and artificial digestant extractable concentration of Cd in the blanched edible parts to assess the risk from oral intake. The experimental results identified that the biochar increased the soil pH and decreased Cd concentration in the roots and shoots of tested vegetables compared with the control. As some chemical forms of Cd in the vegetables were leached out from tissues during cooking, using total Cd in the vegetables over-estimated the dose of Cd absorbed by the human body. Hence, the bioaccessibility of Cd through eating vegetables can be used to predict accurately the health risk of Cd intake, especially under the biochar-amended soil.

Thin-layer capping is an environment-compatible technique for in situ sediment remediation, reducing contaminants released from the solid phases to overlying water. The main approach is to allow the sediment left in place but decreasing further contamination from resuspension of contaminants by the capping layer [8]. Ch'ng investigated mercury (Hg) removal efficiency of iron sulfide (FeS), sulfurized activated carbon (SAC), and raw activated carbon (AC) sorbents influenced by salinity and dissolved organic matter (DOM), and the efficiency of these sorbents as thin layer caps on the remediation of Hg-contaminated estuary sediment to decrease the risk of release [9]. They elucidated that FeS on Hg removal was not significantly affected by salinity levels and maintained with high removal efficiency. The Hg removal efficiency of AC and SAC increased as salinity increased. However, the Hg removal by sorbents decreased with the addition of DOM at different salinity levels. To cope with highly complex conditions in sediment, mixed capping with multiple materials was further performed by Ou et al. [10]. They selected kaolinite, carbon black (CB), iron sulfide (FeS), hydroxyapatite (HAP), and oyster shell powder (OSP) as mixed active caps to retain nickel (Ni), chromium (Cr), copper (Cu), zinc (Zn), and Hg released from freshwater sediment by column experiments. The HAP and OSP showed the highest removal efficiencies towards Ni, Cr, Cu, and Zn, with CB taking the third place. However, the FeS and CB played a more significant role in Hg removal, corresponding to the findings by Ch'ng et al. [9].

The mobility of heavy metals in aquatic environments by desorption from sediments into the surface water is controlled by many biological and chemical factors, making the surface water a major intermediate source of toxic metals in benthic sediments. Aljahdali et al. determined concentrations of heavy metals in sediments and the freshwater mollusc *Bellamya unicolor*, pollution indices, and antioxidant enzyme activities in *Bellamya unicolor* across the five sites in the River Kaduna, Nigeria to further evaluate the risk assessment of heavy metals [11]. They found that a significantly positive correlation between metal concentration and antioxidants catalase and superoxide dismutase was established, supporting the potential ecological risk as a result of heavy metals pollution in the River Kaduna. Martín et al. evaluated the water quality of Boque River in Colombia contaminated by gold mining drainage by bioassays (*Lactuca sativa*, *Hydra attenuata*, and *Daphnia magna*), mutagenicity (Ames test), and microbiological assays, in addition to physiochemical parameters such as pH, heavy metals, Hg, and cyanide [12]. They found Hg, Cd, and cyanide exceeded the permitted concentrations in Colombia and *D. magna* showed sensitivity and *L. sativa* showed inhibition and excessive growth in the analyzed water. The presence of bacteria and coliphages in the water indicated a health risk to inhabitants. The mutagenic index showed the possibility of mutations in the population consuming this type of water. Additionally, bioassays played as an alert system when concentrations of contaminants cannot be analytically detected. In addition to conventional contaminants, emerging contaminants such as antibiotics have received great concerns in the environment worldwide. Yang et al. examined the degradation of antibiotics in the sewage sludge from a wastewater treatment plant by antibiotic-degrading bacteria under aerobic and anaerobic conditions [13]. Four antibiotic-degrading bacterial strains, SF1 (*Pseudmonas* sp.), A12 (*Pseudmonas* sp.), strains B (*Bacillus* sp.), and SANA (*Clostridium* sp.), were isolated, identified, and tested in their study. The experiments indicated the addition of SF1 and A12 under aerobic conditions and the addition of B and SANA under anaerobic conditions increased the biodegradation of antibiotics in the sludge. Moreover, twenty-four reported antibiotics-degrading bacterial genera were identified to have the possible potential for the removal of antibiotics including oxytetracycline (OTC), tetracycline (TC), chlortetracycline (CTC), amoxicillin (AMO), sulfamethazine (SMZ), sulfamethoxazole (SMX), and sulfadimethoxine (SDM) in the sludge.

3. Conclusions

The seven papers in this SI provide valuable results in the topics of soils, sediments, and water contamination according to the consideration of ecological and health risk. They also point out open questions and possible research in the future. Biochar application can benefit both soil conservation and contamination, but further research should be conducted to investigate whether these positive effects can be extended to the field scale. Similar to biochar, scale-up design will be helpful for thin-layer capping in in situ sediment by using mixed active amendments. Both physiochemical analysis and bioassays mutually supported the evaluation results of river water quality. However, we need better approaches and policies of management to prevent further contamination from the discharge of untreated industrial and domestic waste into this aquatic ecosystem. The use of microorganisms to eliminate antibiotics is a promising strategy, but the future work should verify the biodegradation ability of antibiotic-degradation bacteria in the wastewater treatment plant.

Funding: This research received no external funding.

Conflicts of Interest: The author declares no conflict of interest.

References

1. Adhikari, K.; Hartemink, A.E. Linking soils to ecosystem services—A global review. *Geoderma* **2016**, *262*, 101–111. [CrossRef]
2. Hou, D.; Bolan, N.S.; Tsang, D.C.W.; Kirkham, M.B.; O'Connor, D. Sustainable soil use and management: An interdisciplinary and systematic approach. *Sci. Total Environ.* **2020**, *729*, 138961. [CrossRef] [PubMed]

3. Chang, Y.T.; Hsi, H.C.; Hseu, Z.Y.; Jheng, S.L. Chemical stabilization of cadmium in acidic soil using alkaline agronomic and industrial byproducts. *J. Environ. Sci. Health A* **2013**, *48*, 1748–1759. [CrossRef] [PubMed]
4. Kumpiene, J.; Lagerkvist, A.; Maurice, C. Stabilization of As, Cr, Cu, Pb and Zn in soil using amendments—A review. *Waste Manag.* **2008**, *28*, 215–225. [CrossRef] [PubMed]
5. Hamon, R.E.; McLaughlin, M.J.; Cozens, G. Mechanisms of attenuation of metal availability in in-situ remediation treatments. *Environ. Sci. Technol.* **2002**, *36*, 3991–3996. [CrossRef] [PubMed]
6. Kuo, Y.L.; Lee, C.H.; Jien, S.H. Reduction of nutrient leaching potential in coarse-textured soil by using biochar. *Water* **2020**, *12*, 2012. [CrossRef]
7. Chen, K.S.; Pai, C.Y.; Lai, H.Y. Amendment of husk biochar on accumulation and chemical form of cadmium in lettuce and pak-choi grown in contaminated soil. *Water* **2020**, *12*, 868. [CrossRef]
8. Zhang, C.; Zhu, M.Y.; Zeng, G.M.; Yu, Z.G.; Cui, F.; Yang, Z.Z.; Shen, L.Q. Active capping technology: A new environmental remediation of contaminated sediment. *Environ. Sci. Pollut. Res.* **2016**, *23*, 4370–4386. [CrossRef] [PubMed]
9. Ch'ng, B.L.; Hsu, C.J.; Ting, Y.; Wang, Y.L.; Chen, C.; Chang, T.C.; Hsi, H.C. Aqueous mercury removel with carbonaceous and iron sulfide sorbents and their applicability as thin-layer caps in mercury-contaminated estuary sediment. *Water* **2020**, *12*, 1991. [CrossRef]
10. Ou, M.Y.; Ting, Y.; Ch'ng, B.L.; Chen, C.; Cheng, Y.H.; Chang, T.C.; Hsi, H.C. Using mixed active capping to remediate multiple potential toxic metal contaminated sediment for reducing environmental risk. *Water* **2020**, *12*, 1886. [CrossRef]
11. Aljahdali, M.O.; Alhassan, A.B. Metallic pollution and the use of antioxidant enzymes as biomarkers in *Bellamya unicolor* (Olivier, 1804) (Gastropoda: Bellamyinae). *Water* **2020**, *12*, 202. [CrossRef]
12. Martín, A.; Arias, J.; Lópes, J.; Santos, L.; Venegas, C.; Duarte, M.; Ortiz-Ardila, A.; de Parra, N.; Campos, C.; Zambrano, C.C. Evaluation of the effect of gold mining on the water quality in Monterrey, Bolívar (Colombia). *Water* **2020**, *12*, 2523. [CrossRef]
13. Yang, C.W.; Liu, C.; Chang, B.V. Biodegradation of amoxicillin, tetracyclines and sulfonamides in wastewater sludge. *Water* **2020**, *12*, 2147. [CrossRef]

Publisher's Note: MDPI stays neutral with regard to jurisdictional claims in published maps and institutional affiliations.

Article

Reduction of Nutrient Leaching Potential in Coarse-Textured Soil by Using Biochar

Yu-Lin Kuo [1], Chia-Hisng Lee [2] and Shih-Hao Jien [3,*]

[1] Department of Civil Engineering, National Pingtung University of Science and Technology, Pingtung 91201, Taiwan; q3489505@gmail.com

[2] Center for Sustainability Science, Academia Sinica, Taibei 11529, Taiwan; d91623402@ntu.edu.tw

[3] Department of Soil and Water Conservation, National Pingtung University of Science and Technology, Pingtung 91201, Taiwan

* Correspondence: shjien@mail.npust.edu.tw; Tel.: +886-8-7740358; Fax: +886-8-7740373

Received: 4 June 2020; Accepted: 13 July 2020; Published: 15 July 2020

Abstract: *Background*: Loss of nutrients and organic carbon (OC) through leaching or erosion may degrade soil and water quality, which in turn could lead to food insecurity. Adding biochar to soil can effectively improve soil stability, therefore, evaluating the effects of biochar on OC and nutrient retention and leaching is critical. *Methods*: We conducted a 42-day column leaching experiment by using sandy loam soil samples mixed with 2% of biochar pyrolyzed from Honduran mahogany (*Swietenia macrophylla*) wood sawdust at 300 °C (WB300) and 600 °C (WB600) and a control sample. Leaching was achieved by flushing the soil column on day 4 and every week during the 42-day experiment and adding a water volume for each flushing equivalent to the field water capacity. *Results*: Biochar application increased the final soil pH and OC, NH_4^+-N, NO_3^--N, available P concentrations but not exchangeable K concentrations. In particular, WB600 exhibited superior performance in alleviating soil acidification; WB300 engendered high NO_3^--N concentrations. Biochar application effectively retained water in soil and inhibited the leaching of the aforementioned nutrients and dissolved OC. WB300 reduced NH_4^+-N and K leaching by 30%, and WB600 reduced P leaching by 68%. *Conclusions*: Biochar application can improve nutrient retention and reduce the leaching potential of soils and connected water bodies.

Keywords: biochar; organic carbon; nutrients; leaching; nitrogen; phosphorus; potassium

1. Introduction

Nutrients (nitrogen, phosphorus, and potassium) and soil organic carbon (SOC) are critical components of a healthy soil, which is the foundation of a strong food system [1]. Tropical ecosystems are particularly susceptible to the loss of nutrients through soil erosion or leaching processes [2]. Intense rainfall in tropical or subtropical areas results in the leaching of fertilizer containing N, P, and K from soil bodies. Nutrient leaching could diminish soil fertility, accelerate soil acidification, increase fertilizer costs for farmers, and reduce crop yields [3]. The deposition of leached nutrients into water bodies adversely affects aquatic environments because of potential risks such as eutrophication. Leaching of N and P and agricultural runoffs are among the leading contributors to non-point source (NPS) pollution, which has a detrimental effect on drinking and ground water, aquatic habitats, and other water resources. Agricultural runoffs often contain several contaminants, including nutrients, pesticides, pathogens, sediment, salts, trace metals, and other substances, which contribute to biological oxygen demand [4]. Moreover, SOC, comprising nutrient and soil biota, leaches out over time [5], which could aggravate nutrient losses and water pollution. An enormous quantity of fertilizers must be applied to counter the dwindling fertility of agricultural soil.

Biochar is a solid bioresource obtained through the pyrolysis of organic waste. Residues from agricultural and forestry production processes are suitable raw materials for the production of high-quality biochar [6,7]. Biochar is a porous substance containing high levels of carbon and various functional groups. Accordingly, the addition of biochar to agricultural soil has emerged as a feasible strategy to enhance soil water retention capacity [8–10], soil quality [11–14], soil organic matter stability and nutrient retention [15,16], organic carbon (OC) sequestration [17], and greenhouse gases emission reduction [18–21]. Furthermore, biochar can affect soil microbial properties, including microbial activity [22] and microbial diversity [23]. However, the interactions between biochar and microbial properties in soil are not fully understood [24]. The application of biochar to soil could increase soil fertility and crop productivity by reducing leaching or even supplying nutrients [25–27]. However, the effects of biochar on nutrient leaching and OC retention has been reported to vary with the applied biochar pyrolysis temperature, raw material, and soil type [28,29]. Biochar produced from secondary forest residues could reduce fertilizer leaching and increase plant growth and nutrition [26]. Furthermore, the addition of biochar produced from hardwood to a typical Midwestern agricultural soil in the United States considerably reduced the leaching of total N and P by 11% and 69%, respectively [3]. Yao et al. [29] reported that the effect of biochar on nutrient retention and release varied with the nutrient and biochar type.

In this study, we conducted a 42-day column leaching experiment by using loamy sand soil samples that were obtained from a tropical/subtropical area and treated with two types of wood dust biochar pyrolyzed at 300 and 600 °C. The objective of this study was to determine the effects of biochar application on water, nutrient, and OC retention and leaching from the observed soil. The results are expected to be valuable for assessing the potential of biochar for the retention and immobilization of nutrients in soils and inhibition of water body contamination.

2. Materials and Methods

2.1. Collection of Soil Samples and Preparation of Biochars

Surface soil samples (0–15 cm) were collected from a field in Pingtung, Southern Taiwan (22°31′57.9″ N 120°33′38.1″ E). As of April 2016, pineapple (*Ananas comosus* (L.) Merr.) was the dominant crop on this land. The soil samples were air-dried, sieved through a 2-mm screen, and stored at room temperature. The biochar used in this study comprised Honduran mahogany (*Swietenia macrophylla*) wood sawdust obtained from the Department of Wood Design, National Pingtung University of Science and Technology. Two biochar materials were used in this study, namely WB300 and WB600, that were produced at pyrolysis temperatures of 300 and 600 °C, respectively. The biochar used in this study was supplied by the Industrial Technology Research Institute (ITRI) of Taiwan. Before being charred, the wood sawdust was dried at 60 °C for 24 h to < 10% moisture and cut to a particle size of 2 cm. For pyrolysis, the samples were placed in a tubular furnace (ITRI, Tainan, Taiwan) equipped with a corundum tube (diameter, 32 mm; length, 700 mm) and a N_2 purging mechanism (flow rate, 1 L/min) to ensure an oxygen-free atmosphere. Heat treatments were performed at temperatures of 300 and 600 °C, with the heating rate being 5 °C min^{-1}. The temperature was maintained for 2 h before cooling to an ambient temperature under an N_2 flow. After the pyrolysis, the biochar materials were ground to pass through a 2-mm sieve, followed by homogenization through stirring.

2.2. Preparations of Leaching Column

Similar to the procedures applied by Lo [30], the biochar materials were thoroughly mixed with the collected sandy loam soil at application rates of 0% (Control, 0 tons ha^{-1}) and 2% (40 tons ha^{-1}) w/w for Bok choy (*Brassica rapa chinensis*) cultivation in Taiwan. Briefly, nutrient solutions of ammonium sulfate ($(NH_4)_2SO_4$), calcium dihydrogen phosphate ($Ca(H_2PO_4)_2$), and potassium chloride (KCl) were added to the soil at application rates of 2076, 227, and 191 kg ha^{-1}, respectively (approximately 220, 30, and 100 kg ha^{-1} for N, P, and K, respectively). The fertilizers were dissolved in deionized (DI) water and then mixed thoroughly with the soil samples. The volume of the nutrient solution applied was 60% of the water retention capacity of the treated soil samples. The treatment samples are outlined as follows: (1) control, comprising soil only (CK); (2) WB300, comprising soil to which 2% of the biochar pyrolyzed at 300 °C was added; and (3) WB600, comprising the soil to which 2% of the biochar pyrolyzed at 600 °C was added. A leaching experiment was conducted for each treatment in three replicates. As illustrated in Figure 1, a soil column with an internal diameter of 20.6 cm was constructed. The column was composed of two polyvinyl chloride (PVC) tubings of equal length, which were connected through a PVC fitting with a 5-cm interval. A nylon mesh (1 mm^2) with filter paper (Whatman grade no. 42) above was placed between the joints to separate the soil in the upper part from the quartz sand (~2 mm in diameter) filled in the center PVC fitting. At a soil depth of 15 cm, the volume of the soil column was approximately 5000 cm^3. All the columns were packed with the tested soil samples to obtain an initial bulk density of 1.2 g cm^{-3}.

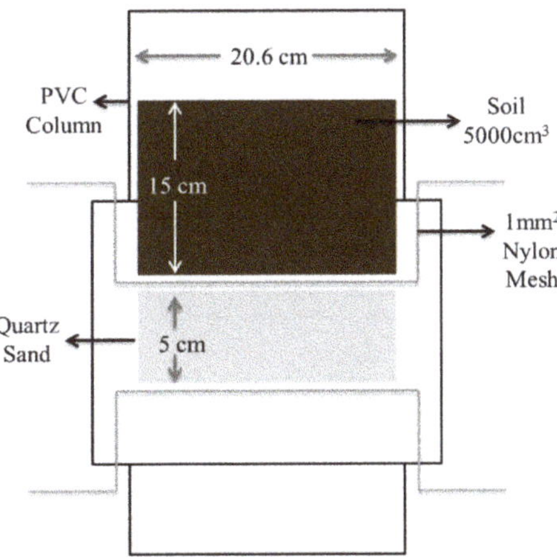

Figure 1. Schematic of the soil column constructed for the leaching experiment.

2.3. *Soil Column Incubation and Leaching*

The soil columns were subjected to a 42-day incubation process conducted at room temperature (25–28 °C) and humidity (60%–80%) with repeated leaching in order to investigate the effects of biochar application on (1) the physicochemical properties of, (2) the hydraulic properties of, and (3) nutrient retention and leaching from the soils. Short-term duration of incubation period was chosen based on Yoo et al. (2013) [31], where the leaching experiment was finished within 60 days. Likewise, based on our previous studies, the variation of chemical and physical properties [13] and dynamic changes of N and P [7] after biochar addition might occur and finish within 8 weeks, therefore, a short-term experiment period (42 days) was selected in this study. Table 1 lists the analysed items.

According to the soil porosity and volume determined for the studied soil column, we used a leaching volume of 700 mL for each flushing process. A fine sieve was placed above the columns to minimize water disturbance to the soil surfaces during flushing. Throughout the experimental period, all columns were leached seven times (on days 4, 7, 14, 21, 28, 35, and 42) using DI water. The leachates were collected using 1000-mL measuring cylinders, and the volumes of the leachates were recorded. Leachates were then subjected to chemical analyses. After the final leaching event, the soil of each column was collected, air-dried, and ground to pass through a 2-mm sieve before further chemical analysis.

Table 1. Analysis items and relevant abbreviations for the leachates and soils.

Properties		Leachate	Soil
Leachate Volume	V_L	✓	
pH	pH		✓
Bulk Density	D_B		✓
Organic Carbon	OC	✓	✓
Ammonium Nitrogen	$NH_4{}^+$-N	✓	✓
Nitrate-Nitrogen	$NO_3{}^-$-N	✓	✓
Available Phosphorus	Ava. P	✓	✓
Exchangeable Potassium	Ex. K	✓	✓

2.4. Analytical Methods

The bulk density (D_B) was determined using the core method [32]. The pH values of the soil samples and biochar materials mixed with DI water (1:2.5 and 1:20 *w/v*, respectively) were determined using a Horiba F-74 BW meter [33]. We performed electrical conductivity measurements on saturated paste extracts of the soil samples by using a Horiba F-74 BW meter [34]. The soil particle size distribution was determined using the pipette method [35]. Cation exchange capacity (CEC) was determined using the ammonium acetate method (pH 7.0) [36]. Exchangeable K was extracted using 1 mol L^{-1} NH_4OAc (1:10 *w/v* for the soil samples; 1:20 *w/v* for the biochar materials), and the extract was analyzed through atomic absorption spectrometry (Z-2300, Hitachi, Tokyo, Japan). The OC concentration was determined through wet oxidation [37]. Available P was determined using the Bray P-1 extract test [38]. Inorganic N was extracted using 2 M KCl (1:10 *w/v*), and the concentrations of $NH_4{}^+$-N and $NO_3{}^-$-N were determined though steam distillation conducted using MgO and Devarda's alloy [39]. The microscale structure of the biochar materials was characterized through optical microscopy using reflected light, followed by scanning electron microscopy (SEM; Hitachi, S-3000N, Japan). A backscattered electron image representing the mean atomic abundance in a black-and-white image was observed on the surface of the samples coated with Au. The C components of biochar horizons were examined through solid-state CPMAS ^{13}C nuclear magnetic resonance (DSX 400-MHz solid-state NMR, Bruker, Karlsruhe, Germany). Data acquisition was executed under the following conditions: spectrometer frequency, 100.46 MHz; spinning speed, 7000 Hz; contact time, 1 ms; and pulse delay time, 1 s. We determined the total signal intensity and the proportion contributed by each C functional group by integrating the spectra in the chemical-shift region: 0–50 ppm (aliphatic C), 50–110 ppm (O-alkyl-C), 110–165 ppm (aromatic C), and 165–190 ppm (carboxyl C). Methoxyl C contributed a wide shoulder between 50 and 60 ppm within the O-alkyl-C range (Alpha-T, Bruker).

2.5. Statistical Analysis

Data were analyzed using IBM SPSS Statistics 22 for Windows (IBM Corp., Armonk, NY, USA). Data sets were subjected to mean separation analysis using one-way analysis of variance, with significance being set to a p value of 0.05. The differences between mean values under different treatments were identified using Duncan's test.

3. Results

3.1. Properties of the Soil and Biochar Materials

Table 2 lists the properties of the soil samples and biochar materials. The texture of the studied soil was sandy loam; the soil was determined to have a neutral pH and low OC content. The porosity, bulk density, and particle density were in the normal ranges for the coarse-textured soil samples. The soil used in this study was sourced from an intensively cultivated field with high human input, which may result in high nutrient concentrations. The pH values of WB300 and WB600 (Honduran mahogany wood sawdust pyrolyzed at 300 and 600 °C) were 6.5 (neutral) and 10.4 (alkaline), respectively. The OC content in WB300 was 6.8%, which was higher than that in WB600 (2.0%). By contrast, the total carbon content was 69% in WB300, which was lower than that in WB600 (79.5%). These results indicate that WB600 contained a higher level of inorganic carbon than W300 did. The higher pyrolysis temperature reduced the concentrations of oxygen, nitrogen, ammonium–nitrogen, nitrate–nitrogen, available phosphorus, and exchangeable potassium in the biochar materials. Figure 2 depicts SEM images of both biochar materials. WB300 exhibited coarser pores than WB600 did but had a lower number of pores for the same volume. Because of its more porous structure—signifying a larger surface area—WB600 could have distinct effects on the physicochemical properties of soil and groundwater when compared with WB300.

Table 2. Properties of the studied soil and Honduran mahogany (*Swietenia macrophylla*) wood sawdust biochar samples pyrolyzed at 300 °C (WB300) and 600 °C (WB600).

Properties	Soil	Wood Biochar (WB)	
		300 °C (WB300)	600 °C (WB600)
pH	6.2	6.5	10.4
EC (dS m^{-1})	0.35	-	-
Sand (%)	72.5	-	-
Silt (%)	18.1	-	-
Clay (%)	9.4	-	-
Texture	SL	-	-
D_B (g cm^{-3})	1.44	-	-
D_P (g cm^{-3})	2.69	-	-
Porosity (%)	42.8	-	-
CEC (cmol(+)/kg)	10.2	55.1	20.4
OC (%)	0.33	6.8	2.0
TC (%)	-	69.0	79.5
H (%)	-	4.5	2.9
O (%)	-	24.7	14.6
N (%)	-	0.92	0.82
H/C	-	0.06	0.04
O/C	-	0.36	0.18
NH_4^+-N (mg/kg)	82.4	56.7	41.7
NO_3^--N (mg/kg)	131	548	341
Ava. P (mg/kg)	6.63	5.29	3.30
Ex. K (mg/kg)	259	449	323

EC: electrical conductivity; SL: sandy loam; D_B: bulk density; D_P: particle density; OC: organic carbon; TC: total carbon; H: hydrogen; O: oxygen; N: nitrogen; NH_4^+-N: ammonium-nitrogen; NO_3^--N: nitrate-nitrogen; Ava. P: available phosphorous; Ex. K: exchangeable potassium; -: Not determined.

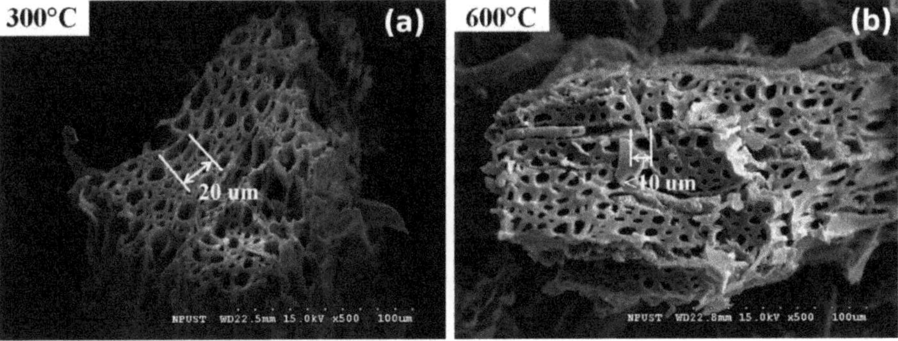

Figure 2. Scanning electron microscope (SEM) images of Honduran mahogany (*Swietenia macrophylla*) wood sawdust biochar pyrolyzed at (**a**) 300 °C and (**b**) 600 °C.

Figure 3 illustrates the functional groups of C within the structures of the WB300 and WB600. O-alkyl-C is the major C group in the natural composition of Honduran mahogany. By pyrolyzing at 300 °C, the WB300 consisted of more aromatic-C, less O-alkyl-C, more alkyl-C, and more carboxylic-C than the raw wood dust. At 600 °C, the pyrolysis process resulted in the predominating aromatic-C in the WB600, and the other C groups became less observable. The results of the physical and chemical properties of the biochars as affected by pyrolysis temperature are consistent with previous studies [7,13,40,41].

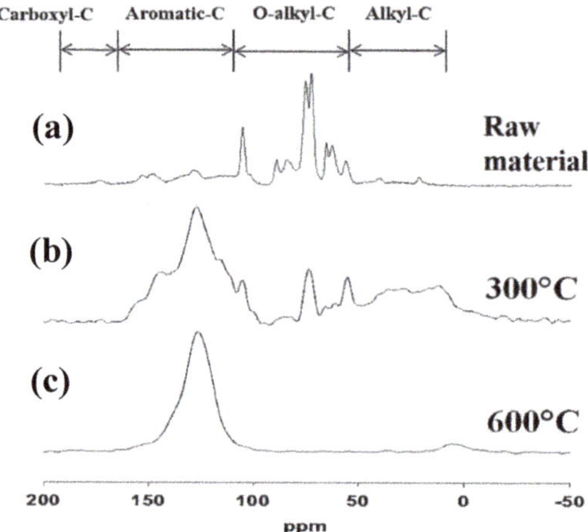

Figure 3. Solid-state ^{13}C cross-polarization magic-angle spinning nuclear magnetic resonance spectra for Honduran mahogany (*Swietenia macrophylla*) wood sawdust (**a**) and its biochar materials pyrolyzed at 300 °C (**b**) and 600 °C (**c**).

3.2. Soil Physicochemical Properties

Table 3 shows the major soil properties under different treatments before and after the 42-days experiment. The pH value of the untreated soil (before mixing with fertilizer) was 6.8, as shown in Table 2. After fertilization, the pH value under control dropped to 6.1. The pH of soil treated with WB300 was ~6.2, whereas WB600 had the highest pH value of 6.5. At day 42, the soil pH values of both biochar treatments were significantly higher than that of the control (pH 4.6), as shown in Table 3. The WB600 treated soil still revealed the highest pH value (5.8) at day 42 among all treatments. Although the soil under WB300 treatment had the similar pH value (6.2) with the control on day 0, it revealed a higher pH than the control at day 42.

The D_B values observed on day 42 for all the treated samples were lower than the initial D_B of 1.20 g cm^{-3} (achieved when the soil columns were packed). The control exhibited the highest D_B (1.11 g cm^{-3}); the WB600-treated sample had the lowest value (1.05 g cm^{-3}). However, the differences in D_B between the treated samples were not significant ($p = 0.05$) (Table 3). Because biochar typically contains low levels of OC, the SOC levels observed for all treatments were low (0.21%–0.46%). The WB300-treated sample had the highest SOC content levels on both day 0 and day 42 (0.46% and 0.39%, respectively), indicating an SOC content loss of only 0.7% throughout the experiment (Table 3). The SOC levels observed for the WB600-treated sample did not differ significantly from that observed for the control (Table 3).

The NH_4^+-N concentration did not differ significantly between any of the treated samples on day 0 (Table 3). On day 42, the NH_4^+-N concentration in all treated samples decreased drastically from approximately 205 mg kg^{-1} to less than 6.5% of the initial concentrations. On day 42, the control had the lowest concentration (5.87 mg kg^{-1}), and the WB300- and WB600-treated samples had significantly higher concentrations (12.7 mg kg^{-1}). The NO_3^--N and inorganic N concentrations in the treated samples exhibited similar trends to the NH_4^+-N concentrations. On day 42, all treated samples exhibited considerably lower NO_3^--N concentrations when compared with the initial concentrations; the NO_3^--N concentrations were high in the samples treated with the two biochar materials, particularly the WB300-treated sample.

Table 3. The Soil physicochemical properties on day 0 and day 42 (n = 3).

Properties	Day	Treatments		
		CK	WB300	WB600
pH	0	6.1 ± 0.1 [a]	6.2 ± 0.1 [a]	6.5 ± 0.1 [b]
	42	4.6 ± 0.1 [a]	5.2 ± 0.2 [b]	5.8 ± 0.1 [c]
D_B (g cm^{-3})	42	1.11 ± 0.24 [a]	1.09 ± 0.11 [a]	1.05 ± 0.10 [a]
SOC (%)	0	0.33 ± 0.05 [a]	0.46 ± 0.03 [b]	0.36 ± 0.05 [a]
	42	0.21 ± 0.05 [a]	0.39 ± 0.05 [b]	0.31 ± 0.07 [a,b]
NH_4^+-N (mg kg^{-1})	0	206 ± 6.28 [a]	205 ± 6.16 [a]	205 ± 6.35 [a]
	42	5.87 ± 1.19 [a]	12.7 ± 1.15 [b]	12.7 ± 1.21 [b]
NO_3^--N (mg kg^{-1})	0	131 ± 8.35 [a]	138 ± 8.98 [a]	135 ± 8.14 [a]
	42	7.52 ± 1.14 [a]	33.0 ± 1.31 [c]	28.0 ± 3.63 [b]
Inorganic N (mg kg^{-1})	0	336 ± 8.61 [a]	343 ± 9.60 [a]	339 ± 8.38 [a]
	42	13.4 ± 0.82 [a]	45.7 ± 0.11 [c]	40.7 ± 3.42 [b]
Ava. P (mg kg^{-1})	0	19.6 ± 0.27 [a]	19.6 ± 0.26 [a]	19.8 ± 0.28 [a]
	42	4.08 ± 0.45 [a]	5.10 ± 0.28 [b]	5.36 ± 0.75 [b]
Ex. K (mg kg^{-1})	0	488 ± 46.2 [a]	457 ± 30.6 [a]	497 ± 45.2 [a]
	42	302 ± 3.02 [a]	315 ± 0.24 [b]	351 ± 0.13 [c]

D_B: bulk density; SOC: soil organic carbon; NH_4^+-N: ammonium–nitrogen; NO_3^--N: nitrate–nitrogen; N: nitrogen; Ava. P: available phosphorous; Ex. K: exchangeable potassium. The values followed by the same superscript letters within a row are not significantly different ($p > 0.05$) between relevant treatments.

On day 0, the Ava. P concentrations did not differ significantly between the three treated soil samples (19.6–19.8 mg kg^{-1}). On day 42, the Ava. P concentration decreased to 4.08, 5.1, and 5.36 mg kg^{-1} in the control, WB300-treated, and WB600-treated samples, respectively. On day 0, the Ex. K concentrations in the control, WB300-treated, and WB600-treated samples were 302, 315, and 351 mg kg^{-1}, respectively. After the experiment, the Ex. K concentrations increased to 457–488 mg kg^{-1} in all treated samples and did not differ significantly between the samples.

3.3. Properties of Leachate

Both biochar-treated samples exhibited significantly smaller leachate volumes than that of the control for each flushing event (Table 4; Figure 4). On day 4 (the day of the first flushing event), the soil column with the control sample had a leachate volume of 530 mL, and both WB300- and WB600-treated samples retained approximately 150 mL more water than the control did (i.e., the leachate volume decreased by 28%). At the end of the experiment, the cumulative leachate volumes observed for the WB300- and WB600-treated samples were lower than that observed for the control by 9.2% and 13.7%, respectively.

Figure 5 displays the cumulative level of dissolved OC (DOC) in the leachate. This was highest in the control and lowest in the WB600-treated sample after the experiment (188 and 154 mg, respectively). After 42 days of incubation, the level of DOC leached from the soil column decreased by 6.50% and 20.0% in the WB300- and WB600-treated samples, respectively, compared with the control. The biochar materials contained low levels of OC. Accordingly, biochar introduces a negligible level of DOC into soils.

Table 4. Volume of the leachate from the soil columns after each flushing with DI water (n = 3).

Treatments	Volume of the Leachate (mL)						
	Incubation Time (Days)						
	4	7	14	21	28	35	42
CK	530 ± 45 ^{b}	539 ± 55	579 ± 39	598 ± 287	625 ± 21	624 ± 82	633 ± 08 ^{b}
WB300	378 ± 63 ^{a}	459 ± 53	539 ± 23	577 ± 18	600 ± 22	598 ± 08	597 ± 11 ^{a}
WB600	377 ± 05 ^{a}	426 ± 55	488 ± 110	525 ± 64	570 ± 05	598 ± 10	578 ± 24 ^{a}

The values followed by the same superscript letters within a column are not significantly different ($p > 0.05$) between the relevant treatments.

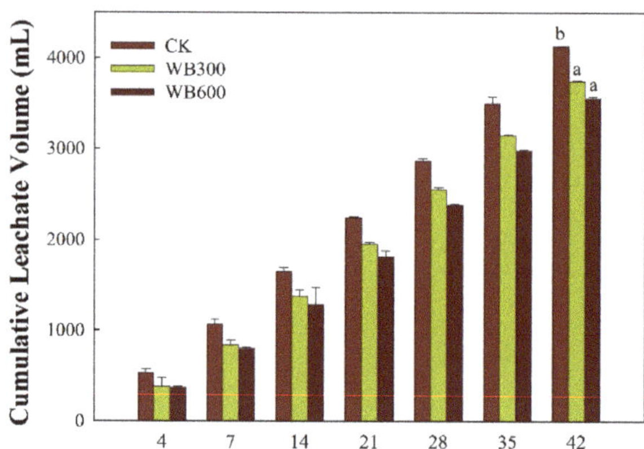

Figure 4. Cumulative leachate volume (V_L) of treated samples (n = 3). Different letters above the bars for day 42 indicate significant differences between the relevant treated samples ($p < 0.05$).

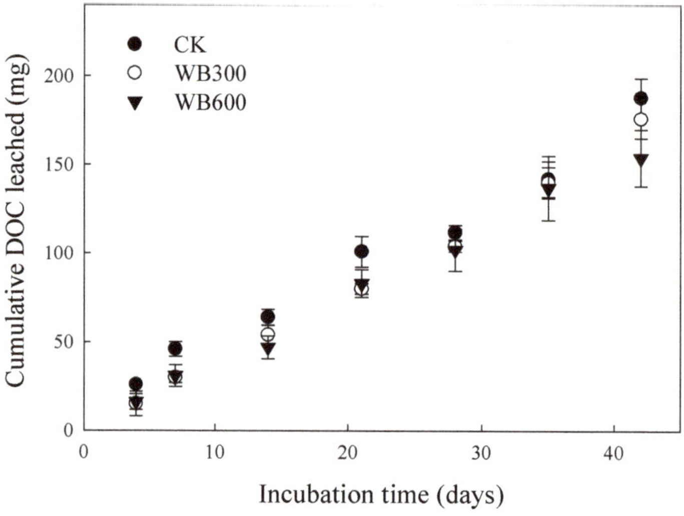

Figure 5. Cumulative concentration of dissolved organic carbon (DOC) for different treated samples (n = 3).

The cumulative quantities of NH_4^+-N and NO_3^--N leached from the soil columns are illustrated in Figure 6. WB300 remarkably reduced NH_4^+-N leaching by 30.5% relative to the control (69.6 mg). Although the inhibitory effect of WB600 on NH4+-N leaching was relatively weak, it still reduced the total quantity of NH_4^+-N leached from the soil by 10.6%, which was approximately one-third of that observed for the WC300-treated sample. The WB300- and WB600-treatments reduced the quantities of NO_3^--N leached from the soil samples by 13.8% and 16.4%, respectively, compared with the control (83.9 mg).

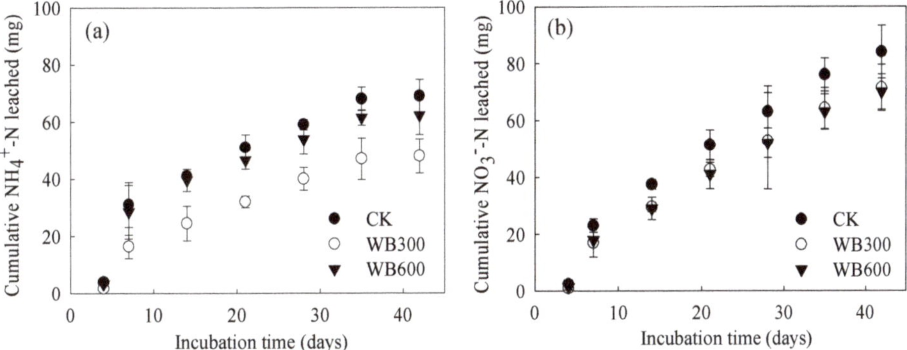

Figure 6. Cumulative quantities of (**a**) NH_4^+-N and (**b**) NO_3^--N leached from the soil columns (n = 3).

The cumulative quantity of inorganic N (summation of the quantities of NH_4^+-N and NO_3^--N) leached from the soil samples subjected to the different treatments differed significantly (Figure 7). The control exhibited the highest quantity of inorganic N leached from the soil (154 mg), and the WB300-treated sample exhibited the lowest quantity (33.3% lower than the control). Furthermore, WB600 treatment decreased the quantity of inorganic N leached from the soil by 13.7% only.

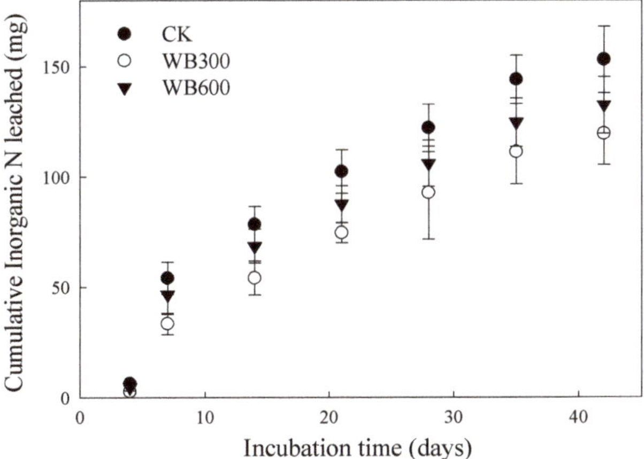

Figure 7. Cumulative quantity of inorganic N leached from the soil columns (n = 3).

Figure 8 displays the cumulative quantities of P leached from the soil columns. The total quantity of P leached from the WB600-treated sample decreased significantly (68.0%) compared with that from the control (12.2 mg). For the WB300-treated sample, the total quantity of P leached from the soil decreased by 45.2%. Compared with the control, the WB300 and WB600 treatments reduced the total quantities of P leached from the soil by 29.71% and 7.70% (156 and 210 mg leached), respectively (Figure 9).

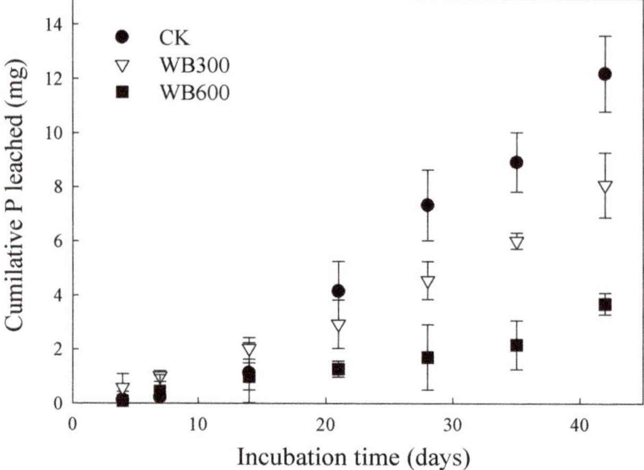

Figure 8. Cumulative quantity of phosphorus leached from the soil columns (n = 3).

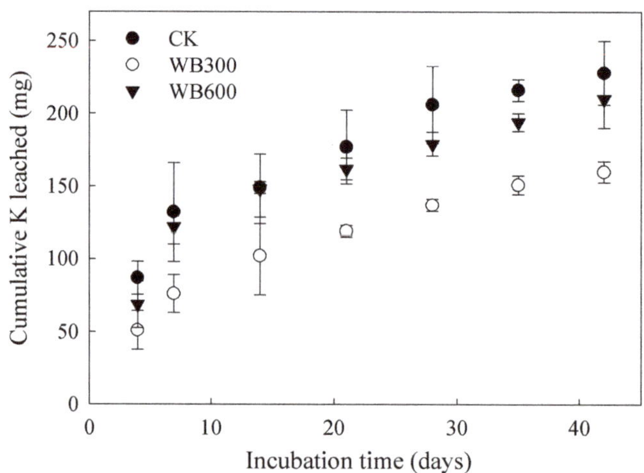

Figure 9. Cumulative quantity of potassium leached from the soil columns (n = 3).

4. Discussion

Compared with the original soil (control on day 0), the pH value of the fertilized soil decreased from 6.8 to 6.1, which could be attributed to the acidic properties of the two fertilizers, namely $(NH_4)_2SO_4$ and $Ca(H_2PO_4)_2$. The pH value of the control decreased to 4.6 on day 42, and those of the WB300- and WB600-treated samples were considerably higher (Table 3). The results indicate that both WB300 and WB600 could alleviate soil acidification. When biochar undergoes pyrolysis at a higher temperature, it generally has a higher pH [42]. Singh et al. [43] revealed that the $CaCO_3$ equivalence of biochar increased with the pyrolysis temperature. Accordingly, the application of biochar could engender liming effects. For biochar, a higher application rate or higher pyrolysis temperature could increase the pH or alkalinity of the biochar-treated soil [11]. The alleviation effects of biochar on soil acidification could contribute to the retardation of the movement of several nutrients and pollutants from soils to groundwater and lower water bodies.

Applying biochar could decrease the D_B of a soil sample [13]. However, this phenomenon was not observed in this study; a possible reason is that the reorganization of the soil structure for mixing the materials and packing the soil columns neutralized the soil properties. A similar condition can be achieved in the field through intensive tilling. For a less- disturbed sandy loam soil, the application of biochar could improve the soil structure, increase the infiltration rate, and reduce runoff water and soil erosion, thus improving soil and water conservation [7,13,41]. The volumes of leachates increased with repeated leaching for all treatments in this study. Although the differences in leachate volume between the control and biochar-treated samples decreased gradually in this study, the biochar-treated samples still retained a significantly higher amount of water in the soil columns on day 42 (Table 4). The WB300 and WB600 treatments reduced the level of water loss by 9.2% and 13.7%, respectively. Our results demonstrate that the biochar materials, particularly WB600, exhibited a strong ability to conserve water in the soil samples when applied at a rate of 2%.

The OC contents in both of the biochars were much lower than that of other organic materials commonly used in farmland. The biochar applications displayed significantly affect the SOC in the WB300 treatment only in this study (Table 3). Accordingly, the effects of biochar on DOC leaching are the results of the sorption of organic carbon onto the biochar, either within the pores of the biochar or onto the external biochar surface [7]. The biochar-applied soils retained more water than the control (Figure 4), as illustrated earlier, which could also contribute to DOC retention since DOC move down and leach out from the soil column along with the soil water. Our results illustrated that the highest amounts of DOC and leachate were found in CK indicating biochar addition could effectively retain DOC in the soils. However, the highest DOC concentration (47.0 mg/L, DOC amount/leachate volume) was found in the leachate of WB300 treatment indicated that biochar might release solube C components into soil solution, particularly in biochar pyrolized with low temperatures [7,44]. The results implicated that soluable C components onto the biochar itself might increase DOC concentration in output water (runoff or eluate from soil pedon) duing rainfall events in the biochar-amended soils. The WB300 treatment revealed the efficiency of 9.2% for water retention and the lower efficiency of 6.5% for DOC retention (Figures 4 and 5). Under WB600 treatment, the retention efficiencies were 13.7% for water and a higher of 20.0% for DOC. This result indicates that WB600 had a stronger affinity to DOC than WB300. Kasozi et al. [45] reported that the organic matter sorption onto biochar surfaces is kinetically limited by slow diffusion into the subnanometer-sized pores dominating biochar surfaces. The various organo-mineral interactions lead to aggregations of soil and organic materials, which stabilizes both soil structure and the carbon compounds within the aggregates. Furthermore, the increase in the diversity and density of carbon groups within WB300 biochar may result in the slightly increased SOC but did not prevent the leaching of DOC as effective as the WB600 biochar.

Although the NH_4^+-N and NO_3^--N concentrations in the soil columns were low on day 42 (Table 3), the WB300- and WB600-treated samples, especially the WB300-treated sample, exhibited higher inorganic N concentrations than the control did. These results could be attributed to the high surface area and diverse functional groups, such as carboxyl C, O-alkyl-C, and alkyl C, of WB300 (Figure 3). Obvious higher CEC was also fould in WB300 than in WB600 in this study indicating that more NH_4^+-N and K could be retained in the soil treated with WB300, and which were demonstrated by our results in Figures 6 and 9 [46,47]. Addtionally, improvement in soil physical properties such as promotion of soil aggregation and increasing of water holding capacity might also bring positive effects in nutrient leaching. Yoo et al. [31] suggested that increasing formation of aggregates by biochar addition could effectively promoted retention of NO_3^-. Furthermore, they also indicated that increased water holding capacity after biochar addition was also a factor to reduce N leaching.

Overall, both biochars can effectively reduce N leaching and provide a potential N source of nutrient delivery to plants [48]. Agricultural non-point source (NPS) pollution is the leading source of water quality impacts to rivers and lakes. Nitrogen from fertilizers, manure, waste and ammonia turns into nitrite and nitrate. High levels of these toxins deplete waters of oxygen, killing all of the animals and fish. Nitrates also soak into the ground and end up in drinking water. Health problems can occur as a result of this and they contribute to methemeglopbinemia or blue baby syndrome which causes death in infants. Base on our results, application biochar might be a useful managememt practice to reduce NPS pollution in watersheds, particularly in tropical/subtropical climate regions. In this study, we measured the N losses through leaching. The N losses through denitrification and ammonia volatilization, which may form NH_3, N_2, or N_2O, were not accessed. As the soil was pH 6.8 and the soil columns were nearly water-saturated during the experiment, we used acidic fertilizer, $(NH_4)_2SO_4$ and $Ca(H_2PO_4)_2$ intentionally to minimize ammonium volatilization.

Applying the biochar materials engendered only a slight increase in the available P in the soil samples after the leaching experiment (Table 3). Both biochar materials, particularly WB600, could retain soil P, according to the leaching results (Figure 8). This effect also resulted in the higher concentrations of Ava. P in the soil samples after the experiment, as mentioned. P tends to precipitate with Fe under acidic conditions. However, the application of the biochar materials increased the soil pH, which could enhance the release of P. Therefore, the biochar materials were likely to have contained numerous bonding sites or other co-precipitation elements, resulting in considerably higher P retention efficiency levels compared with their water retention efficiency levels. During soil pH 4 to 6, increasing of soil pH might unlock P "adsorbed on soluble and hydrous Fe/Al oxides" into an available form, however, the unlocked P might be adsorbed again onto biochar to form a potential available form. Our results was consistence with the resuld of Laird et al. [3] who indicated that biochar addition could effectively reduce leaching of dissolved P in the soil column due to adsorption of orthophosphate and adsorption of organic P compounds by biochar.

The Ex. K concentration was higher in the biochar-treated soil samples, particularly the WB600-treated sample, than in the control at the beginning of the experiment; this could be attributed to the high mobility of K in plant ash. A higher pyrolysis temperature may result in a higher concentration of soluble K in biochar. However, the Ex. K concentrations did not differ significantly between the samples on day 42. Both biochar materials considerably reduced the total quantity of K leached from the soil samples. The efficiency levels of WB300 and WB600 in inhibiting K leaching were 29.7% and 7.7%, respectively. WB600 inhibited K leaching mainly through holding the soil solutions, and WB300 possibly had additional K sorption mechanisms mainly related to C functional groups.

Our results indicated that both WB300 and WB600 could effectively reduce the leaching of soil water, DOC, NH_4^+-N, NO_3^--N, P, and K. After incubation, not only nutrient concentration but also leachate volume were found lower in biochar-treated soil compared with those in control (Table 4). Reharding the strong water retention capacity of the biochar-treated soils, two possible reasons were speculated, which were (1) high degree of evaporation (about 30 °C in average during summer in southern Taiwan) for the soil column after water adding; therefore, the biochar treatments might still effectively retain water after water addition; (2) the volume of adding water (700 mL) to the soil column at each date still did not match the pore volume (~2000 cm^3) of soil column; therefore, the biochar treatments might still effectively retain water after water addition at day 7 and other dates. Nutrient leaching could be inhibited through increased water retention. Moreover, the inhibitory effect of the biochar materials on the mineralization of organic N, in terms of physical protection of organic matter [7,40], can reduce the quantity of nutrients released and thus reduce subsequent leaching. Our results reveal that WB300 exhibited a high NH_4^+-N and K (predominately cations) retention efficiency, and WB600 exhibited a high water, DOC, and P (predominately anions) retention efficiency. These results indicate that WB300 was negatively charged, which was most likely due to the distribution of various carbon functional groups. Overall, the biochar-induced retention of soil water, DOC, and nutrients could be considered to positively affect nutrient and water conservation and to improve soil quality. Reducing the leaching of water, DOC, and nutrients from soils could conserve groundwater and connected water bodies. Therefore, biochar application can benefit both soil and water conservation. Further research should be conducted to investigate whether these positive effects can be extended to the field and to downstream water bodies at the catchment level.

5. Conclusions

The results of this study demonstrate that the incorporation of Honduran mahogany (*Swietenia macrophylla*) wood sawdust biochar into sandy loam soil samples could improve soil health by increasing the capacity of the soil to retain nutrients and reduce nutrient leaching. The biochar materials applied in this study, particularly WB600, alleviated soil acidification. The incorporation of the biochar materials considerably increased SOC, inorganic N (NH_4^+-N and NO_3^--N), and ava. P concentrations in the soil samples. The biochar application did not significantly affect D_B or Ex. K in the soil. Furthermore, the total volume of leachates and cumulative quantity of DOC, inorganic N, P, and K leached from the soil samples decreased significantly in the biochar-treated samples. WB300 engendered the least quantities of NH_4^+-N and K, both of which are cations. Thus, WB300 could be more negatively charged than WB600 because of its inhibitory effects on cation leaching. By contrast, WB600 was likely to have a higher density of P-bonding sites, which resulted in a stronger inhibitory effect on P leaching. Although the ability to retain various nutrients in soil differed with pyrolysis temperatures, both biochar materials effectively contributed to the conservation of groundwater and subsequent downstream water bodies. Increased retention of these nutrients in soils can increase the probability of absorption by plant roots, thereby decreasing the risk of leaching into rivers or groundwater reservoirs. In future studies, the on-site effects of biochar application on underground and water bodies should be determined in terms of eutrophication and potential pollution, especially for intensively fertilized cropped fields.

Author Contributions: Conceptualization and methodology, S.-H.J.; sample analysis, Y.-L.K.; writing—original draft preparation, S.-H.J. and Y.-L.K.; writing—review and editing, S.-H.J. and C.-H.L. All authors have read and agreed to the published version of the manuscript.

Funding: This research was funded by the Ministry of Science and Technology of the R.O.C. (Grant number MOST 105-2628-B-020-001-MY2).

Acknowledgments: We would like to thank the students of the Soil Survey and Conservation Laboratory, Department of Soil and Water Conservation, National Pingtung University of Science and Technology, for their assistance during the column leaching experiment and analysis.

Conflicts of Interest: The authors declare no conflicts of interest.

References

1. Robertson, M. *Sustainability Principles and Practice*, 2nd ed.; Taylor & Francis: London, UK, 2017.
2. Jordan, C.F.; Herrera, R. Tropical rain forests: Are nutrients really critical? *Am. Nat.* **1981**, *117*, 167–180. [CrossRef]
3. Laird, D.; Fleming, P.; Wang, B.; Horton, R.; Karlen, D. Biochar impact on nutrient leaching from a Midwestern agricultural soil. *Geoderma* **2010**, *158*, 436–442. [CrossRef]
4. O'Green, A.T.; Budd, R.; Gan, J.; Manynard, J.J.; Parikh, S.J.; Dahlgren, R.A. Mitigating nonpoint source pollution in agriculture with constructed and restored wetlands. *Adv. Agron.* **2010**, *108*, 1–76.
5. Lal, R. Tillage and agricultural sustainability. *Soil Tillage Res.* **1991**, *20*, 133–146. [CrossRef]
6. Lu, K.; Yang, X.; Shen, J.; Robinson, B.; Huang, H.; Liu, D.; Bolan, N.; Pei, J.; Wang, H. Effect of bamboo and rice straw biochars on the bioavailability of Cd, Cu, Pb and Zn to Sedum plumbizincicola. *Agric. Ecosyst. Environ.* **2014**, *191*, 124–132. [CrossRef]
7. Jien, S.H.; Wang, C.C.; Lee, C.H.; Lee, T.Y. Stabilization of organic matter by biochar application in compost-amended soils with contrasting pH values and textures. *Sustainability* **2015**, *7*, 13317–13333. [CrossRef]
8. Busscher, W.J.; Novak, J.M.; Evans, D.E.; Watts, D.W.; Niandou, M.; Ahmedna, M. Influence of pecan biochar on physical properties of a Norfolk loamy sand. *Soil Sci.* **2010**, *175*, 10–14. [CrossRef]
9. Kammann, C.I.; Linsel, S.; Gößling, J.W.; Koyro, H.W. Influence of biochar on drought tolerance of Chenopodium quinoa Willd and on soil–plant relations. *Plant Soil* **2011**, *345*, 195–210. [CrossRef]

10. Basso, A.S.; Miguez, F.E.; Laird, D.A.; Horton, R.; Westgate, M. Assessing potential of biochar for increasing water-holding capacity of sandy soils. *Gcb Bioenergy* **2013**, *5*, 132–143. [CrossRef]
11. Yuan, J.H.; Xu, R.K.; Zhang, H. The forms of alkalis in the biochar produced from crop residues at different temperatures. *Bioresour. Technol.* **2011**, *102*, 3488–3497. [CrossRef]
12. Vaccari, F.; Baronti, S.; Lugato, E.; Genesio, L.; Castaldi, S.; Fornasier, F.; Miglietta, F. Biochar as a strategy to sequester carbon and increase yield in durum wheat. *Eur. J. Agron.* **2011**, *34*, 231–238. [CrossRef]
13. Jien, S.H.; Wang, C.S. Effects of biochar on soil properties and erosion potential in a highly weathered soil. *Catena* **2013**, *110*, 225–233. [CrossRef]
14. Zhao, X.; Wang, J.; Xu, H.; Zhou, C.; Wang, S.; Xing, G. Effects of crop-straw biochar on crop growth and soil fertility over a wheat-millet rotation in soils of China. *Soil Use Manag.* **2014**, *30*, 311–319. [CrossRef]
15. Clough, T.J.; Condron, L.M.; Kammann, C.; Müller, C. A review of biochar and soil nitrogen dynamics. *Agronomy* **2013**, *3*, 275–293. [CrossRef]
16. Ventura, M.; Sorrenti, G.; Panzacchi, P.; George, E.; Tonon, G. Biochar reduces short-term nitrate leaching from a horizon in an apple orchard. *J. Environ. Qual.* **2013**, *42*, 76–82. [CrossRef]
17. Laird, D.A. The charcoal vision: A win–win–win scenario for simultaneously producing bioenergy, permanently sequestering carbon, while improving soil and water quality. *Agron. J.* **2008**, *100*, 178–181.
18. Gaunt, J.L.; Lehmann, L. Energy balance and emissions associated with biochar sequestration and pyrolysis bioenergy production. *Environ. Sci. Technol.* **2008**, *42*, 4152–4158. [CrossRef]
19. Woolf, D.; Amonette, J.E.; Street-Perrott, F.A.; Lehmann, J.; Joseph, S. Sustainable biochar to mitigate global climate change. *Nat. Commun.* **2010**, *1*, 1–9. [CrossRef]
20. Zheng, J.; Stewart, C.E.; Cotrufo, M.F. Biochar and nitrogen fertilizer alters soil nitrogen dynamics and greenhouse gas fluxes from two temperate soils. *J. Environ. Qual.* **2012**, *41*, 1361–1370. [CrossRef]
21. Stewart, C.E.; Zheng, J.; Botte, J.; Cotrufo, M.F. Co-generated fast pyrolysis biochar mitigates green-house gas emissions and increases carbon sequestration in temperate soils. *GCB Bioenergy* **2013**, *5*, 153–164. [CrossRef]
22. Warnock, D.D.; Lehmann, J.; Kuyper, T.W.; Rillig, M.C. Mycorrhizal responses to biochar in soil–concepts and mechanisms. *Plant Soil* **2007**, *300*, 9–20. [CrossRef]
23. Jin, H. *Characterization of Microbial Life Colonizing Biochar and Biochar-Amended Soils*; Cornell University: New York, NY, USA, 2010.
24. Lehmann, J.; Rillig, M.C.; Thies, J.; Masiello, C.A.; Hockaday, W.C.; Crowley, D. Biochar effects on soil biota—A review. *Soil Biol. Biochem.* **2011**, *43*, 1812–1836. [CrossRef]
25. Glaser, B.; Lehmann, J.; Zech, W. Ameliorating physical and chemical properties of highly weathered soils in the tropics with charcoal—A review. *Biol. Fertil. Soils* **2002**, *35*, 219–230. [CrossRef]
26. Lehmann, J.; da Silva, J.P., Jr.; Steiner, C.; Nehls, T.; Zech, W.; Glaser, B. Nutrient availability and leaching in an archaeological Anthrosol and a Ferralsol of the Central Amazon basin: Fertilizer, manure and charcoal amendments. *Plant Soil* **2003**, *249*, 343–357. [CrossRef]
27. Major, J.; Rondon, M.; Molina, D.; Riha, S.J.; Lehmann, J. Maize yield and nutrition during 4 years after biochar application to a Colombian savanna oxisol. *Plant Soil* **2010**, *333*, 117–128. [CrossRef]
28. Mukherjee, A.; Zimmerman, A.R. Organic carbon and nutrient release from a range of laboratory-produced biochars and biochar–soil mixtures. *Geoderma* **2013**, *193*, 122–130. [CrossRef]
29. Yao, Y.; Gao, B.; Zhang, M.; Inyang, M.; Zimmerman, A.R. Effect of biochar amendment on sorption and leaching of nitrate, ammonium, and phosphate in a sandy soil. *Chemosphere* **2012**, *89*, 1467–1471. [CrossRef]
30. Lo, C.H. *Manual of Fertilizer Application*, 6th ed.; Agriculture and Food Agency, Council of Agriculture, Executive Yuan: Taipei, Taiwan, 2005.
31. Yoo, G.Y.; Kim, H.J.; Chen, J.J.; Kim, Y.S. Effects of biochar addition on nitrogen leaching and soil structure following fertilizer application to rice paddy soil. *Soil Sci. Soc. Am. J.* **2013**, *78*, 852–860. [CrossRef]
32. Blake, G.R.; Hartge, K.H. Bulk density. In *Methods of Soil Analysis, Part 1. Physical and Mineralogical Methods*; Klut, A., Ed.; ASA and SSSA: Madison, WI, USA, 1986; Agronomy monograph No. 9; pp. 363–375.
33. McLean, E. Soil pH and lime requirement. In *Methods of Soil Analysis. Part 2. Chemical and Microbiological Properties*; Page, A.L., Miller, R.H., Keeney, D.R., Eds.; ASA and SSSA: Madison, WI, USA, 1982; pp. 199–224.
34. Rhoades, J. 1982. Cation exchange capacity. In *Methods of Soil Analysis. Part 2. Chemical and Microbiological Properties*; Page, A.L., Miller, R.H., Keeney, D.R., Eds.; ASA and SSSA: Madison, WI, USA, 1982; pp. 149–157.

35. Gee, G.W.; Bauder, J.W.; Klute, A. Particle-size analysis. In *Methods of Soil Analysis. Part 1. Physical and Mineralogical Methods*; Klut, A., Ed.; ASA and SSSA: Madison, WI, USA, 1986; Agronomy monograph No. 9; pp. 383–411.
36. Sumner, M.E.; Miller, W.P. Cation exchange capacity and exchange coefficients. In *Methods of Soil Analysis: Soil Science Society of America Book Series 5 Part 3—Chemical Methods*; Sparks, D.L., Page, A.L., Helmke, P.A., Loeppert, R.H., Eds.; ASA and SSSA: Madison, WI, USA, 1996; pp. 1201–1229.
37. Nelson, D.W.; Sommers, L.E.; Sparks, D.; Page, A.; Helmke, P.; Loeppert, R.; Soltanpour, P.; Tabatabai, M.; Johnston, C.; Sumner, M. Total carbon, organic carbon, and organic matter. In *Methods of Soil Analysis: Soil Science Society of America Book Series 5 Part 3—Chemical Methods*; Sparks, D.L., Page, A.L., Helmke, P.A., Loeppert, R.H., Eds.; ASA and SSSA: Madison, WI, USA, 1996; pp. 961–1010.
38. Bray, R.H.; Kurtz, L. Determination of total, organic, and available forms of phosphorus in soils. *Soil Sci.* **1945**, *59*, 39–46. [CrossRef]
39. Mulvaney, R.L. Nitrogen-Inorganic forms. In *Methods of Soil Analysis: Soil Science Society of America Book Series 5 Part 3—Chemical Methods*; Sparks, D.L., Page, A.L., Helmke, P.A., Loeppert, R.H., Eds.; ASA and SSSA: Madison, WI, USA, 1996; pp. 1123–1184.
40. Jien, S.H.; Chen, W.C.; Ok, Y.S.; Awad, Y.M.; Liao, C.S. Short-term biochar application induced variation of C and N mineralization in a compost-added rural soil. *Environ. Sci. Pollut. Res.* **2018**. [CrossRef]
41. Lee, C.H.; Wang, C.C.; Lin, H.H.; Lee, S.S.; Tsang, D.C.W.; Jien, S.H.; Ok, Y.S. In situ biochar application conserves water and nutrients while simultaneously mitigating the erosion of an Fe-oxide-enriched tropical soil. *Sci. Total Environ.* **2018**, *619*, 665–671. [CrossRef] [PubMed]
42. Cantrell, K.B.; Hunt, P.G.; Uchimiya, M.; Novak, J.M.; Ro, K.S. Impact of pyrolysis temperature and manure source on physicochemical characteristics of biochar. *Bioresour. Technol.* **2012**, *107*, 419–428. [CrossRef] [PubMed]
43. Singh, B.; Singh, B.P.; Cowie, A.L. Characterisation and evaluation of biochars for their application as a soil amendment. *Soil Res.* **2010**, *48*, 516–525. [CrossRef]
44. Mukherjee, A.; Zimmerman, A.R.; Harris, W. Surface chemistry variations among a series of laboratory-produced biochars. *Geoderma* **2011**, *163*, 247–255. [CrossRef]
45. Kasozi, G.N.; Zimmerman, A.R.; Kizza, P.N.; Gao, B. Catechol and Humic Acid Sorption onto a Range of LaboratoryProduced Black Carbons (Biochars). *Environ. Sci. Technol.* **2010**, *44*, 6189–6195. [CrossRef] [PubMed]
46. Steiner, C.; Glaser, B.; Teixeira, W.G.; Lehmann, J.; Blum, W.E.H.; Zech, W. Nitrogen retention and plant uptake on a highly weathered central Amazonian Ferralsol amended with compost and charcoal. *J. Plant Nutr. Soil Sci.* **2008**, *171*, 893–899. [CrossRef]
47. Lentz, R.D.; Ippolito, J.A. Biochar and manure affect calcareous soil and corn silage nutrient concentrations and uptake. *J. Environ. Qual.* **2011**, *41*, 1033–1043. [CrossRef]
48. Hagemann, N.; Kammann, C.; Schmidt, H.P.; Kappler, A. Nitrate capture and slow release in biochar amended compost and soil. *PLoS ONE* **2017**, *12*, e171214. [CrossRef]

Article

Amendment of Husk Biochar on Accumulation and Chemical Form of Cadmium in Lettuce and Pak-Choi Grown in Contaminated Soil

Kuei-San Chen [1], Chun-Yu Pai [2] and Hung-Yu Lai [1,3,*]

1 Department of Soil and Environmental Sciences, National Chung Hsing University, Taichung 40227, Taiwan; s104039023@smail.nchu.edu.tw
2 Department of Agronomy, National Taiwan University, Taipei 10617, Taiwan; r08628311@ntu.edu.tw
3 Innovation and Development Center of Sustainable Agriculture, National Chung Hsing University, Taichung 40227, Taiwan
* Correspondence: soil.lai@nchu.edu.tw; Tel.: +886-422-840-373

Received: 11 February 2020; Accepted: 18 March 2020; Published: 20 March 2020

Abstract: (1) Background: Cadmium (Cd) accumulated in vegetables not only affects their growth but can also enter the human body via food chains and lead to various illnesses. Plants can decrease the toxicity by changing the chemical forms of Cd, which include inorganic (F_E), water-soluble (F_W), pectate- and protein-integrated (F_{NaCl}), undissolved phosphate (F_{HAc}), oxalate (F_{HCl}), and residual forms (F_R). Among them, F_E and F_W chemical forms show higher mobility to translocate upward from roots to shoots compared with the others. (2) Methods: Different varieties or cultivars of lettuce and pak-choi were grown in Cd-contaminated soils amended with husk biochar (BC) to replenish nitrogen to the recommended amount and also to raise the soil pH value. (3) Results: More than 73% of the accumulated Cd in the edible organs was compartmentalized in F_E chemical form in both leafy vegetables regardless of treatments. In comparison with control, the application of BC decreased the Cd concentrations and bioconcentration factors in the roots and shoots of two leafy vegetables at different growth periods in general. The chemical form and bioaccessible fraction of Cd in the edible blanching tissues were used to calculate the risk of oral intake. The vegetable-induced hazard quotients of lettuce and pak-choi were acceptable, except for pak-choi grown in control without applying BC.

Keywords: bioaccessibility; cadmium; chemical form; husk biochar; risk assessment

---

1. Introduction

Many illegal factories arbitrarily discharge wastewater into irrigation channels, where farmers unknowingly draw contaminated water for paddy rice cultivation, causing many farmlands in Taiwan to be contaminated with potential toxic elements (PTEs) [1]. If the total concentration of PTEs in the soil is above the control standard stipulated by the Soil and Groundwater Pollution Remediation Act (SGWPR Act) of Taiwan, crops should be removed to prevent human consumption [2]. However, when the total concentration of PTEs in the soil is slightly above the monitoring standard and below the control standard, crop safety must be emphasized, especially concerning the pathway of oral intake of vegetables.

Cadmium (Cd), a non-essential PTE, is one of the most toxic substances to plants and animals because it is easily transferrable from soil to plant [3,4]. The accumulation of Cd in the edible tissues of crops poses serious concerns to the trophic risk of Cd transfer along a food chain [5]. Because of the carcinogenic and adverse effects of Cd on biological processes, the International Agency of Research on Cancer and the United States Environmental Protection Agency classified Cd into Group 1 and

Class B, respectively [6]. Excessive concentration of Cd damaged mesophyll cells and epidermal cells in wheats [7] and resulted in the reduction of photosynthesis and chlorophyll content [8,9]. Once the roots uptake PTEs, plants can decrease the toxicity of PTEs by cell wall deposition and vacuolar compartmentation [10–13]. Cell walls are mainly composed of pectin, protein, cellulose, and semi-cellulose [14], and functional groups in cell walls could restrict PTEs into the cytoplasm, which prevents the protoplast from being poisoned [15].

The chemical forms can be of help in understanding the tolerance and mechanisms of PTE detoxification in plants. These chemical forms include inorganic (F_E), water-soluble (F_W), pectate- and protein-integrated (F_{NaCl}), undissolved phosphate (F_{HAc}), oxalate (F_{HCl}), and residual (F_R) forms [16,17]. Among them, F_E and F_W have high mobility and thus are easily translocated to other plant organs [18]. Many studies revealed that F_{NaCl} was related to adjusting Cd stress in ramie, cabbage and American pokeweed [18–20]. There was also a positive linear correlation between Cd concentration in the F_{NaCl} and F_{HAc} forms, and that in the shoot, which revealed that chemical forms can be used to predict the accumulation of Cd in the shoot [21].

Many soil amendments (e.g., compost, phosphate fertilizer, and organic matter) have been used for remediation of PTE-contaminated soil [22,23]. For instance, Liu et al. [24] found that the co-application of vermicompost and selenium could alleviate the Cd accumulation in rice. The presence of organic matter redistributed the PTE to less-available forms and thus ameliorated its toxicity to plants [25]. The solubility of PTE is high in an acidic pH, whereas addition of lime and biochar (BC) can increase the soil pH and thus can decrease the bioavailability [26–28]. Biochars are porous, low density, and carbon-rich solid products from the pyrolysis of waste biomass. Xu et al. [29] revealed that adding BCs to contaminated paddy soil improved the transformation of Cd from the acid-soluble fraction to the oxidizing and residual fractions. BCs also had a great capacity to reduce the percentage of inorganic and water-soluble fractions in lettuce roots grown in Cd-contaminated soil [30]. Additionally, BCs mainly increased undissolved cadmium phosphate and thus increased the Cd accumulation in pokeweed root [31].

In this study, different varieties or cultivars of lettuce and pak-choi were grown in soil dominantly contaminated with Cd. Husk BC was applied to replenish nitrogen to the recommended amount and to decrease the soil acidity. The objective of this study was to understand the effect of the application of husk BC on the accumulation, translocation, and chemical forms of Cd in two leafy vegetables. In addition, the vegetable-induced hazard quotient (HQ_v) was calculated via the chemical form and artificial digestant extractable concentration of Cd in the blanched edible parts to assess the risk from oral intake.

2. Materials and Methods

Experimental soil samples were collected from the surface layer (0–30 cm) of a Cd-contaminated site in central Taiwan because the soil Cd resulted from the use of irrigation water mainly accumulated in the top 30 cm [32]. Soil samples were air-dried, ground, passed through a 10-mesh stainless steel sieve, and then homogenized before analysis. Their basic properties were then analyzed, including pH [33], electrical conductivity (EC) [34], content of organic carbon (OC) [35], cation exchange capacity (CEC) [36], texture [37], and total concentration of Cd, chromium (Cr), nickel (Ni), and zinc (Zn) [38]. Other soil samples were homogenized and used for a pot experiment. The total concentrations of nitrogen, phosphorus, and potassium in the BC were analyzed in accordance with the methods described by Bremner [39], Kuo [40], and Helmke and Sparks [41], respectively. Two treatments used in this study were CK (control without applying biochar) and BC, which applied husk BC to replenish nitrogen to the recommended amount. For lettuce and pak-choi, 46,920 and 98,000 kg/ha of BC was applied, respectively. In total, 1.0 kg of soil samples or mixture was added to each pot, and seeds of two pak-choi—*Brassica chinensis* L. var. Chinensis (PCC) and *Brassica chinensis* L. cv. Wrinkled leaf (PCW)—and two lettuce—*Lactuca sative* L. cv. Chinese (LSC) and *Lactuca sative* L. var. Sative (LSS)—were sown. The above vegetables were selected because of their importance in human diets [42] and they are the most commonly consumed leafy vegetables in Taiwan [43]. The pots were translocated to a phytotron (25.1 ± 0.4 °C, relative humidity 61.2 ± 5.7%). Soil moistures were determined every two to three days, and remained at 60–80% of water-holding capacity by replenished deionized water.

All leafy vegetables were harvested 35, 42, and 49 days (D35, D42, and D49) after sowing and divided into roots and shoots. At this time, the shoot heights were measured and the chlorophyll content of the largest extended leaf was determined using a Konica Minolta SPAD-502 and recorded as SPAD (soil plant analyzer development) readings. Fresh plant tissues were rinsed with tap water and then deionized water. To remove the adsorbed Cd, the roots were soaked in 20 mM of Na_2-EDTA for 15 min. The Cd compartmentalized in different chemical forms in the shoots was analyzed using a sequential extraction in accordance with the methods described by Lai [21]. Six chemical forms were extracted in the following sequence, with the corresponding agents: inorganic Cd (F_E) extracted by 80% alcohol, water-soluble Cd (F_W) extracted by deionized water, pectate- and protein-integrated Cd (F_{NaCl}) extracted by 1 M NaCl, undissolved Cd phosphate (F_{HAc}) extracted by 2% CH_3COOH, Cd oxalate (F_{HCl}) extracted by 0.6 M HCl, and residual Cd (F_R) digested with aqua regia. All the other plant tissues were oven-dried at 65 °C for 72 h and then weighed (dry weight) before grinding (Rong Tsong Precision Tech. Co., Taichung, Taiwan), and then digested with $HNO_3/HClO_4$ (v/v = 3/1). The Cd concentration in the digestant was determined using a flame atomic absorption spectrophotometer (FAAS, Perkin Elmer AAnalyst 200, Waltham, MA, USA).

Soil samples of different treatments were air-dried, ground, passed through a 10-mesh stainless steel sieve, and then homogenized before analysis. The available concentration of Cd in the soils was extracted with 0.1 N HCl [44] and 0.05 N EDTA [45]. A BCR sequestration extraction based on the European Community Bureau of Reference [46] was also conducted to understand the Cd distribution in the different fractions. These four fractions include the exchangeable fraction (F-I), organic matter-bounding fraction (F-II), oxide-bounding fraction (F-III), and residual fraction (F-IV).

Statistical analysis was performed using the Statistical Package for the Social Sciences (SPSS, Armonk, NY, USA). Analysis of variance (ANOVA) was used to test the effect of different treatments on the different growth exhibitions and soil properties. The least significant difference (LSD) test was used to compare the significant differences of pH, EC, available Cd concentration, Cd distribution in the different fractions, four growth exhibitions, Cd concentration in the roots and shoots, bioconcentration factor (BCF = ratio of root concentration to soil concentration), and transfer factor (TF = shoot concentration to root concentration) between treatments ($p < 0.05$).

3. Results

3.1. Soil Properties

The total concentrations of Cd, Cr, Ni, and Zn were in the ranges of 10–13, 51–60, 126–146, and 61–68 mg/kg, respectively. For the four PTEs analyzed, only Cd and Ni were beyond the control standard (Cd 5 mg/kg, Ni 130 mg/kg) of farmland based on the SGWPR Act of Taiwan and could cause the samples to be regarded as Cd- and Ni-contaminated soils. However, only Cd accumulation in the lettuce and pak-choi was discussed in this paper because of following reasons. First, the total concentration of Cd was beyond not only the Canadian soil quality guideline (1.4 mg/kg) but also the Soil Contamination Warming Limit (4 mg/kg) and Counterplan Limit (12 mg/kg) of Korea for agricultural lands. Moreover, Cd has higher bioaccessibility in comparison with Ni [47] and is also the only one PTE regulated by the Office of the Journal of the European Union [48] among four soil PTEs analyzed in this study. The BC used in this study had very strong alkalinity and high EC. Its pH and EC were 10.54 ± 0.06 and 1.82 ± 0.08 dS/m, respectively. Total concentrations of nitrogen, phosphorus, and potassium in the BC were 0.24 ± 0.05%, 0.61 ± 0.10%, and 1.59 ± 0.09%, respectively. For the four PTEs analyzed in this study, the BC only had 1.40 ± 1.99 and 10.42 ± 2.25 mg/kg of Cd and Zn, respectively, and the total concentrations of Cr and Ni were not detectable.

The application of BC was able to significantly increase the soil pH and EC in comparison with CK. The BC treatments raised or significantly ($p < 0.05$) raised pH from 6.6–6.8 and 6.9–7.0 to 6.8–7.1 and 7.6–8.0 for the lettuce and pak-choi cultivated soils, respectively (Table 1). Because of the high EC of BC, the soil EC also increased or significantly ($p < 0.05$) increased from 0.03–0.08 (CK) to 0.06–0.18 dS/m under the BC treatment.

Table 1. Effect of different treatments on the pH, electrical conductivity, two chemical agents extractable Cd concentrations, and Cd's distribution in the different fractions.

Treatment [2]	Soil Property and Cd Concentration [1]							
	pH	Electrical Conductivity	Extracting Agent		BCR Sequestration Extraction [3]			
			0.1 N HCl	0.05 N EDTA	F-I	F-II	F-III	F-IV
		dS/m	mg/kg		%			
PCC-CK	6.90 ± 0.15b	0.03 ± 0.00b	6.63 ± 0.24a	6.81 ± 0.25a	55.99 ± 0.39b	26.37 ± 0.23c	16.18 ± 0.22b	1.46 ± 0.04b
PCC-BC	7.93 ± 0.18a	0.12 ± 0.02a	6.98 ± 0.61a	7.03 ± 0.38a	64.67 ± 0.31a	26.61 ± 0.07c	7.63 ± 0.10c	1.08 ± 0.03c
PCW-CK	6.90 ± 0.20b	0.05 ± 0.01b	7.23 ± 0.68a	7.01 ± 0.39a	52.57 ± 0.34c	31.00 ± 0.10a	16.44 ± 0.17ab	ND
PCW-BC	7.62 ± 0.46a	0.17 ± 0.06ab	6.91 ± 0.56a	6.27 ± 0.34b	51.93 ± 0.29d	29.91 ± 0.15b	16.56 ± 0.22a	1.60 ± 0.03a
LSC-CK	6.69 ± 0.11a	0.07 ± 0.02a	7.70 ± 0.62a	7.41 ± 0.85a	61.34 ± 0.27a	20.86 ± 0.19c	17.80 ± 0.23b	ND
LSC-BC	6.81 ± 0.16a	0.16 ± 0.08a	7.00 ± 0.64a	6.82 ± 0.56a	37.70 ± 0.39d	45.26 ± 0.11a	17.04 ± 0.22c	ND
LSS-CK	6.77 ± 0.11a	0.05 ± 0.01a	6.90 ± 0.88a	7.62 ± 0.30a	38.90 ± 0.28c	36.51 ± 0.32b	21.84 ± 0.16a	2.76 ± 0.05
LSS-BC	7.02 ± 0.23a	0.06 ± 0.00a	6.55 ± 1.18a	6.64 ± 0.94a	57.32 ± 0.44b	20.88 ± 0.12c	21.80 ± 0.25a	ND

[1] Mean ± standard deviation; ND: not detectable; The same lowercase letter indicates no significant difference between treatments for the same soil property and the same leafy vegetable. [2] PCC, PCW, LSC, and LSS are *Brassica chinensis* L. var. Chinensis, *Brassica chinensis* L. cv. Wrinkled leaf, *Lactuca sative* L. cv. Chinese, and *Lactuca sative* L. var. Sative, respectively; CK: control without applying husk biochar; BC: applying 46,920 and 98,000 kg/ha of husk biochar for lettuce and pak-choi, respectively. [3] F-I, F-II, F-III, and F-IV are exchangeable fraction, organic matter-bounding fraction, oxide-bounding fraction, and residual fraction, respectively.

3.2. *Soil Cd Fraction and Growth Exhibition*

Even through the BC treatment raised the soil pH in comparison with CK, the soil Cd fractions only slightly changed except for those of the lettuce. Regardless of treatments and crops, F-I was the primary fraction of Cd in the soils, and accounted for 37–65% of the total concentration in general (Table 1). In comparison with CK, the BC treatment significantly ($p < 0.05$) changed the Cd in F-I and F-II of lettuce. For LSS treated with BC, the Cd proportion of F-I significantly ($p < 0.05$) increased, but F-II significantly ($p < 0.05$) decreased compared with CK. For LSC treated with BC and in contrast with CK, however, the Cd in the F-I significantly ($p < 0.05$) decreased and F-II significantly ($p < 0.05$) increased. The BC treatment did not significantly change the Ni proportion in the different fractions (detailed data not shown). F-IV was the primary fraction of Ni in general and accounted for 57–63% of the total Ni in the soils. Besides sequence extraction, 0.1 N HCl and 0.05 N EDTA were used to extract the available concentration of Cd and Ni in the soils. Approximately 58–68% and 56–69% of the total Cd in the soils could be extracted by 0.1 N HCl and 0.05 N EDTA, respectively. Because most of the Ni was presented in the F-IV fraction, 0.1 N HCl and 0.05 N EDTA only extracted 10–15% and 9–13% of the total Ni in the soils, respectively. In comparison with CK, the BC treatment did not significantly decrease the two chemical agents' extractable Cd and Ni concentrations in the soils in general.

Compared to CK treatments, the root length, shoot height, shoot fresh weight, and SPAD reading increased under BC treatment in general (Table 2). However, most of the differences were not statistically significant. These results revealed that BC treatment promoted the growth of PCC in general.

Table 2. The growth exhibitions (mean ± standard deviation) of pak-choi and lettuce grown in the contaminated soils amended with or without applying husk biochar [1].

Growth Exhibition	Pak-choi [2]				Lettuce [2]			
	PCC-CK	PCC-BC	PCW-CK	PCW-BC	LSC-CK	LSC-BC	LSS-CK	LSS-BC
Root length (cm)	4.61 ± 1.49a	8.26 ± 3.52a	3.59 ± 0.20a	4.66 ± 0.95a	7.59 ± 1.30a	6.96 ± 0.21a	7.71 ± 0.45a	8.09 ± 1.38a
Shoot height (cm)	10.32 ± 2.75a	12.19 ± 0.67a	12.19 ± 0.94a	11.56 ± 1.79a	8.74 ± 0.38bc	7.58 ± 0.88c	15.47 ± 0.39a	17.99 ± 2.19ab
Shoot fresh weight (g/plant)	0.48 ± 0.07a	1.03 ± 0.12a	0.51 ± 0.28a	0.78 ± 0.56a	0.52 ± 0.09c	0.90 ± 0.23b	0.35 ± 0.08c	1.51 ± 0.22a
SPAD reading [3]	10.48 ± 1.70ab	7.84 ± 0.60c	9.71 ± 0.92bc	11.96 ± 1.14a	14.32 ± 1.97a	14.41 ± 1.74a	12.42 ± 0.50a	14.91 ± 0.30a

[1] The same lowercase letter indicates no significant difference between treatments for the same leafy vegetable and the same growth exhibition. [2] The meanings of abbreviations are the same as in Table 1. [3] SPAD reading: soil plant analyzer development reading.

3.3. Cd Accumulation

The concentrations of PTEs in the lettuce and pak-choi at three different growth periods (D35, D42, and D49) are shown in Figures 1 and 2. The Cd concentrations in the roots of LSC cultivars grown in CK for 35–49 days were in the range of 3.1–11.0 mg/kg, which were 1.4 to 6.4 times higher than that accumulated in the roots of LSS (Figure 1). In the shoots of different growing periods of lettuce, LSC also accumulated higher concentrations of Cd compared with LSS. The Cd concentrations in the shoots of LSC and LSS grown in CK of D35–D49 were in the ranges of 2.1–6.9 and 1.8–3.1 mg/kg, respectively. The LSC accumulated higher concentrations of Cd both in roots and shoots compared with LSS. Relative to LSS, the accumulated Cd concentrations in the edible parts of LSC were 1.7–3.5 times higher, which revealed that LSC can be regarded as a high-Cd-accumulation cultivar of lettuce. The PCW cultivar accumulated higher concentrations of Cd in the roots and shoots in comparison with PCC in general (Figure 2). The accumulated Cd concentrations in the roots of PCW and PCC growing for D35–D49 were in the ranges of 6.6–32.3 and 2.9–12.7 mg/kg, respectively. In the shoots of PCW and PCC, the Cd concentrations of D35–D49 were 3.4–8.9 and 3.4–7.8 mg/kg, respectively.

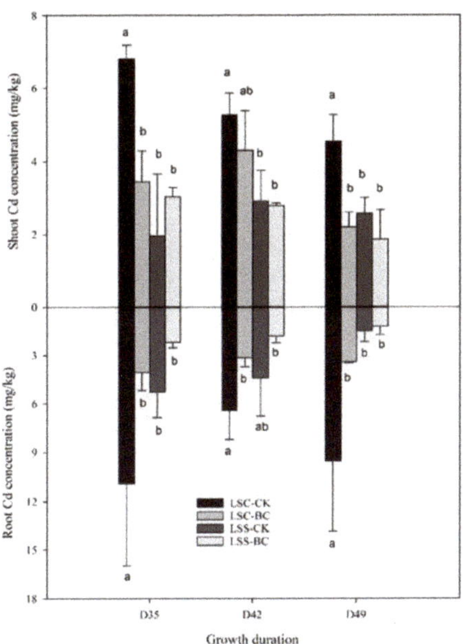

Figure 1. Effect of husk biochar treatment on the Cd accumulation in the roots and shoots of lettuce. The meanings of abbreviations are the same as in Table 1. The same lowercase letter indicates no significant difference between treatments for the same growth period.

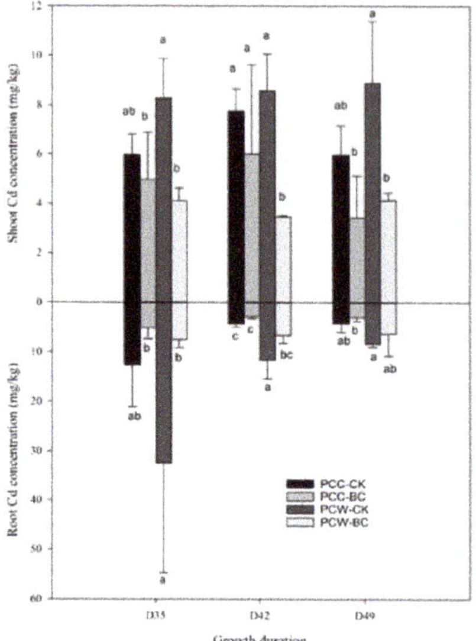

Figure 2. Effect of husk biochar treatment on the Cd accumulation in the roots and shoots of pak-choi. The meanings of abbreviations are the same as in Table 1. The same lowercase letter indicates no significant difference between treatments for the same growth period.

Except for LSS-D35, in comparison with CK, the BC treatments decreased or significantly ($p < 0.05$) decreased the accumulation of Cd in the roots and shoots of lettuce (Figure 1) and pak-choi (Figure 2). This phenomenon was especially true for the LSC variety, which had high accumulating capacity compared with LSS. Relative to CK, the Cd concentrations accumulated in the roots and shoots of LSC under BC treatments decreased 50–64% and 18–52%, respectively, in comparison with CK. Whether treating with BC or not, higher Cd concentrations were found in the roots and shoots of pak-choi compared with lettuce at different growth periods. Under the BC treatments, most of the Cd concentrations in the roots and shoots of pak-choi decreased 24–77% and 16–60%, respectively, in comparison with CK.

3.4. Chemical Form

According to the standard for the Tolerance of Heavy Metals in Plant Origin of Taiwan for leafy vegetables, 0.2 mg/kg·FW, most of the Cd concentrations in the edible parts of lettuce and pak-choi were beyond this standard if the water content of them was 90%. Whether treating with BC or not, Cd was mainly compartmentalized in the F_E chemical form in the shoots of lettuce and pak-choi. Approximately 94–98% and 82–98% of the Cd accumulated in the shoots of lettuce and pak-choi, respectively, were compartmentalized in the F_E at D42–D49 (Figure 3).

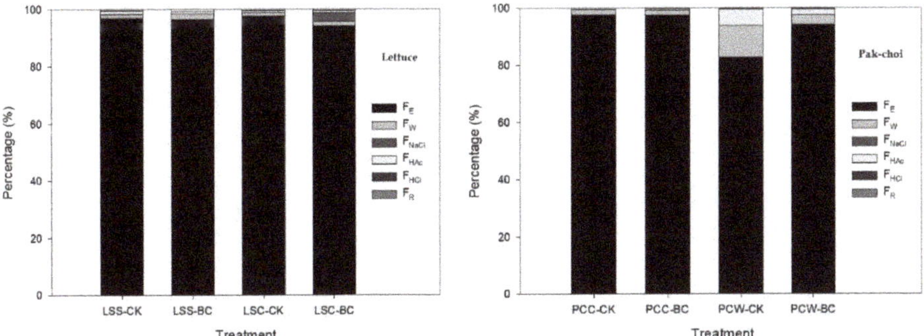

Figure 3. Effect of husk biochar treatment on the chemical form of Cd in the shoots of pak-choi and lettuce. The meanings of treatments are the same as in Table 1. F_E, F_W, F_{NaCl}, F_{HAc}, F_{HCl}, and F_R are inorganic, water-soluble, pectate- and protein-integrated, undissolved phosphate, oxalate, and residual chemical forms, respectively.

4. Discussion

For the three different growing periods, the accumulated Cd in the edible parts of LSC and PCW grown in CK were in the ranges of 4.5–6.9 and 8.2–8.9 mg/kg, respectively (Figures 1 and 2). These values were 1.7–3.5 and 1.1–1.5 times higher than LSS and PCC, respectively. Further results revealed that the accumulating capacity of Cd of LSC and PCW was stronger than LSS and PCC. Moreover, the lettuce and pak-choi accumulated almost constant concentration of Cd in the edible parts regardless of the growing period once planted in the Cd-contaminated soils. The experimental results of this study are in agreement with Lai and Chen [49].

BCF was used to evaluate the accumulation of Cd in the edible parts of two leafy vegetables at D49 (Table 3). BCF_R and BCF_S are the ratio of root and shoot concentration to soil concentration, respectively. The LSC accumulated more Cd than LSS, and the BCF_R of LSC at D49 was 6.1 and 2.6 times higher than LSS grown in CK and BC, respectively. All the BCF_R and BCF_S of lettuce grown in CK were less than 0.9, and there was a 25–68% and 32–58% decrease in BCF_R and BCF_S, respectively, under BC treatment in comparison with CK. Whether applying BC or not, the BCF_S of LSC was 1.1–1.8 times higher than that of LSS. This phenomenon reveals that LSC can accumulate higher Cd concentrations in the edible parts and also shows higher risk through oral consumption than LSS.

The PCW accumulated more Cd than PCC, and the BCF_R of PCW at D49 was 1.9 and 2.3 times higher than PCC grown in CK and BC, respectively. All the BCF_R and BCF_S of pak-choi grown in CK were less than 0.9, and there was a 27–55% and 45–50% decrease in BCF_R and BCF_S, respectively, under BC treatment compared with CK. Even though the BCF_R of PCW was higher than PCC, the BCF_S of PCW grown in different treatments was 57–75% that of PCC. This phenomenon reveals that the Cd accumulated in the roots of PCW was not efficiently upwardly transferred to shoots and thus had lower BCF_S compared with PCC.

Besides BCF, TF was used to evaluate the transfer of Cd from roots to edible parts of lettuce and pak-choi (Table 3). Approximately all the TF values of LSC and PCW, with higher BCF_R and BCF_S under different growth periods and treatments, were below unity in general. For LSS and PCC, however, TF values were higher than LSC and PCC, respectively, and were in the ranges of 0.3–1.8 and 0.4–2.1, respectively. Further results reveal that the Cd uptake by LSC and PCW grown in CK were mainly accumulated in the non-edible parts, regardless of growing periods. However, the TF values of LSS and PCC grown in BC at D42 and D49 reached 1.1–2.1. This result reveals that the accumulated Cd would transfer to shoots more easily than the other two varieties, even grown in BC. Except for LSS-D49, the BC treatment was efficient in increasing the TF of LSS and LSC in comparison with CK, regardless of growth periods. This result revealed that even the BC decreased the accumulated Cd concentration in the roots; on the contrary, BC increased the upward translocation of Cd from roots to shoots in two lettuce varieties because of Cd's high mobility compared with other PTEs.

Table 3. The bioconcentration factor (BCF), transfer factor (TF), average daily dose (ADD), and hazard quotient (HQ) calculated using three different methods [1].

Treatment [2]	BCF_R [3]	BCF_S [3]	TF [3]	Average Daily Dose [4] (ADD_v; μg/kg·BW/day)			Hazard Quotient (HQ_v) [4]		
				ADD_v-TC	ADD_v-CF	ADD_v-BF	HQ_v-TC	HQ_v-CF	HQ_v-BF
PCC-CK	0.40 ± 0.17ab	0.57 ± 0.11a	1.43 ± 0.67a	0.66	0.002	0.15	7.88	0.02	1.73
PCC-BC	0.25 ± 0.06b	0.29 ± 0.14a	1.15 ± 0.64a	0.37	0.001	0.08	4.42	0.02	0.97
PCW-CK	0.78 ± 0.06a	0.82 ± 0.23a	1.06 ± 0.23a	0.60	0.019	0.13	7.16	0.22	1.57
PCW-BC	0.57 ± 0.40ab	0.37 ± 0.03a	0.65 ± 0.49a	0.29	0.004	0.06	3.43	0.05	0.75
LSC-CK	0.89 ± 0.40a	0.42 ± 0.07a	0.47 ± 0.37b	0.34	0.002	0.08	4.07	0.03	0.90
LSC-BC	0.28 ± 0.01b	0.18 ± 0.04b	0.64 ± 0.13b	0.20	0.004	0.04	2.41	0.05	0.53
LSS-CK	0.14 ± 0.06b	0.25 ± 0.04b	1.70 ± 0.49a	0.26	0.002	0.06	3.04	0.03	0.67
LSS-BC	0.11 ± 0.05b	0.17 ± 0.07b	1.53 ± 0.79a	0.29	0.002	0.06	3.46	0.03	0.76

[1] The same lowercase letter indicates no significant difference between treatments for the same leafy vegetable. [2] The meanings of abbreviations are the same as in Table 1. [3] BCF_R: ratio of root conc. to soil conc.; BCF_S: ratio of shoot conc. to soil conc.; TF: ratio of shoot conc. to root conc. [4] ADDv and HQ_v are vegetable-induced ADD and HQ based on the total concentration, chemical form, and bioaccessible fraction of Cd, respectively.

In comparison with CK, the BC treatment did not have significant effects on changing the 0.1 N HCl and 0.05 N EDTA extractable Cd concentrations (Table 1), and also the chemical form of Cd compartmentalized in the shoots of lettuce and pak-choi (detail data not shown). However, the accumulated Cd concentration in the roots and shoots of lettuce and pak-choi grown under BC treatment at different growth periods were decreased or significantly decreased ($p < 0.05$) compared with CK (Figures 1 and 2).

Antoniadis et al. [50] reported that the vegetable-induced average daily dose (ADD_v) and vegetable-induced hazard quotient (HQ_v) can be calculated using Equations (1) and (2), respectively, where C_p is the Cd concentration (mg/kg) in the edible parts of vegetables. The mean individual daily vegetable consumption (MIDVC) in Taiwan during 2013–2016 was 0.133 kg/day based on the Report on the Nutrition and Health Survey and vegetable calorie counts, which can be used to calculate Cd intake daily per person from vegetables. The tolerable daily intake (TDI) of Cd set by the European Food Safety Authority (EFSA) was 0.36 μg/kg·BW/day. Nonetheless, food was the dominant source of Cd exposure of humans and accounts for approximately 90% of the intake [51]. Among all foods, approximately 26% was from vegetables [52], which means that the TDI from vegetables (TDI_v) is 0.084 μg/kg·BW/day.

$$ADD_v = \frac{C_p \times MIDVC}{kg \cdot BW} \quad (1)$$

$$HQ_v = \frac{ADD_v}{TDI_v} \quad (2)$$

Blanching is the most common method in Taiwan for cooking leafy vegetables, and it also decreases the concentration of PTE in thoroughly cooked vegetables. Based on the findings of Lam et al. [47], approximately 50% of the Cd accumulated in the water spinach was leached into boiling water. In this study, three methods based on total concentration (TC), chemical form (CF), and bioaccessible fraction (BF) of Cd in the edible parts of vegetables were used to calculate the HQ_v. The F_E and F_W were considered to have a higher mobility than other chemical forms and were easily leached into boiling water. Therefore, the sum of the proportion of the other four chemical forms, i.e., F_{NaCl}, F_{HAc}, F_{HCl}, and F_R, was used to calculate the HQ_v, coded as HQ_v-CF. Furthermore, approximately 32–55% (average is 44%) of the accumulated Cd in water spinach could be metabolized by in vitro digestive fluids [47], which reveals that approximately 44% of the Cd is bioaccessible and can be absorbed by the human body, coded as HQ_v-BF.

Regardless of treatments, the HQ_V-TC, HQ_V-CF, and HQ_V-BF values of usedleafy vegetables at D49 were in the ranges of 2.4–7.9, 0.01–0.3, and 0.5–1.8, respectively (Table 3). Because more than 73% of the accumulated Cd was compartmentalized in the F_E chemical form, which could be leached out of vegetable tissues during blanching, the HQ_V-CF values were less than 0.3 in general. The application of BC significantly decreased the HQ_V of pak-choi at D49, and the HQ_V-TC, HQ_V-CF, and HQ_V-BF was 20–89% in comparison with the CK. However, BC's effect on the HQ_V of lettuce was contrary to that of pak-choi because the HQ_V increased compared with CK. The HQ_V of lettuce and pak-choi used in this study was lower in comparison with water spinach grown in artificially Cd-spiked soils with a total concentration of 2.8–3.1 mg/kg [53]. According to the calculated results of HQ_V-CF and HQ_V-BF, except for pak-choi grown in CK, oral intake of these four leafy vegetables has a low risk even though the soil Cd concentration was 2 to 3 times beyond the control standard of farmland, i.e., 5 mg/kg, based on the SGWPR Act of Taiwan.

5. Conclusions

Experimental results evidenced that the application of BC was able to increase the soil pH and decrease the accumulation of Cd in the roots and shoots of leafy vegetables used in this study. The BCF also decreased under BC treatment compared with control. However, BC's effect on the upward transfer of Cd from root to shoot was dependent on crop species. Because some of the accumulated Cd in the vegetables will leach out of tissues during cooking, using total concentration of Cd in the vegetables cannot actually reflect the real dose of Cd absorbed by the human body. Based on the HQ_V calculation using the chemical form and bioaccessible fraction of Cd in the edible parts of blanched leafy vegetables, all of the vegetables grown in the study soil had low risk through oral intake, especially under BC treatment.

Author Contributions: Conceptualization and methodology, H.-Y.L.; sample analysis, K.-S.C. and C.-Y.P.; writing—original draft preparation, K.-S.C. and C.-Y.P.; writing—review and editing, H.-Y.L. All authors have read and agreed to the published version of the manuscript.

Funding: This research was funded by the Ministry of Science and Technology of the R.O.C. (Grant number MOST 108-2313-B-005-026) and by the Ministry of Education, Taiwan, R.O.C., under the Higher Education Sprout Project.

Acknowledgments: We would like to thank the students of the Soil Survey and Remediation Laboratory, Department of Soil and Environmental Sciences, National Chung Hsing University, for their assistance during the pot experiment and analysis.

Conflicts of Interest: The authors declare no conflicts of interest.

References

1. Hseu, Z.Y.; Su, S.W.; Lai, H.Y.; Guo, H.Y.; Chen, T.C.; Chen, Z.S. Remediation techniques and heavy metal uptake by different rice varieties in metal-contaminated soils of Taiwan: New aspects for food safety regulation and sustainable agriculture. *Soil Sci. Plant Nutr.* **2010**, *56*, 31–52. [CrossRef]
2. SGWPR Act (Soil and Groundwater Pollution Remediation Act). Available online: https://sgw.epa.gov.tw/en/laws_policy/laws/458db6eb-5602-46b2-9471-745d58078aaf (accessed on 5 March 2020).
3. Di Toppi, L.S.; Gabbrielli, R. Response to cadmium in higher plants. *Environ. Exp. Bot.* **1999**, *41*, 105–130. [CrossRef]
4. Chen, D.; Guo, H.; Li, R.; Li, L.; Pan, G.; Chang, A.; Joseph, S. Low uptake affinity cultivars with biochar to tackle Cd-tainted rice—A field study over four rice seasons in Hunan, China. *Sci. Total Environ.* **2016**, *541*, 1489–1498. [CrossRef]
5. Ali, H.; Khan, E.; Ilahi, I. Environmental chemistry and ecotoxicology of hazardous heavy metals: Environmental persistence, toxicity, and bioaccumulation. *J. Chem.* **2019**, *2019*, 1–14. [CrossRef]
6. Barraza, F.; Schreck, E.; Lévêque, T.; Uzu, G.; López, F.; Ruales, J.; Prunier, J.; Marquet, A.; Maurice, L. Cadmium bioaccumulation and gastric bioaccessibility in cacao: A field study in areas impacted by oil activities in Ecuador. *Environ. Pollut.* **2017**, *229*, 950–963. [CrossRef] [PubMed]

7. Kovacevic, G.; Kastori, R.; Merkulov, L. Dry matter and leaf structure in young wheat plants as affected by cadmium, lead, and nickel. *Biol. Plant.* **1999**, *42*, 119–123. [CrossRef]
8. Liu, Y.T.; Chen, Z.S.; Hong, C.Y. Cadmium-induced physiological response and antioxidant enzyme changes in the novel cadmium accumulator, *Tagetes patula*. *J. Hazard. Mater.* **2011**, *189*, 724–731. [CrossRef]
9. Xu, D.Y.; Chen, Z.F.; Sun, K.; Yan, D.; Kang, M.J.; Zhao, Y. Effect of cadmium on the physiological parameters and the subcellular cadmium localization in the potato (*Solanum tuberosum* L.). *Ecotox. Environ. Safe.* **2013**, *97*, 147–153. [CrossRef]
10. Lasat, M.M.; Baker, A.J.M.; Kochian, L.V. Physiological characterization of root Zn^{2+} absorption and translocation to shoots in Zn hyperaccumulator and nonaccumulator species of Thlaspi. *Plant Physiol.* **1996**, *112*, 1715–1722. [CrossRef]
11. Lasat, M.M.; Fuhrmann, M.; Ebbs, S.D.; Cornish, J.E.; Kochian, L.V. Phytoremediation of a radiocesium-contaminated soil: Evaluation of cesium-137 bioaccumulation in the shoots of tree plant species. *J. Environ. Qual.* **1998**, *7*, 165–169. [CrossRef]
12. Küpper, H.; Zhao, F.J.; McGrath, S.P. Cellular compartmentation of zinc in leaves of the hyperaccumulator *Thlaspi caerulescens*. *Plant Physiol.* **1999**, *119*, 305–311. [CrossRef] [PubMed]
13. Ge, W.; Jiao, Y.Q.; Sun, B.L.; Qin, R.; Jiang, W.S.; Liu, D.H. Cadmium-mediated oxidative stress and ultrastructural changes in root cells of poplar cultivars. *S. Afr. J. Bot.* **2012**, *83*, 98–108. [CrossRef]
14. Haynes, R.J. Ion exchange properties of roots and ionic interactions within root apoplasm: Their role in ion accumulation by plants. *Bot. Rev.* **1980**, *46*, 75–99. [CrossRef]
15. Zhao, Y.F.; Wu, J.F.; Shang, D.R.; Ning, J.S.; Zhai, Y.X.; Shend, X.F.; Ding, H.Y. Subcellular distribution and chemical forms of cadmium in the edible seaweed, *Porphyra yezoensis*. *Food Chem.* **2015**, *168*, 48–54. [CrossRef] [PubMed]
16. Wu, F.B.; Dong, J.; Qian, Q.Q.; Zhang, G.P. Subcellular distribution and chemical form of Cd and Ca-Zn interaction in different barley genotypes. *Chemosphere* **2005**, *60*, 1437–1446. [CrossRef]
17. Su, Y.; Liu, J.L.; Lu, Z.W.; Wang, X.M.; Zhang, Z.; Shi, G.G. Effects of iron deficiency on subcellular distribution and chemical forms of cadmium in peanut roots in relation to its translocation. *Environ. Exp. Bot.* **2014**, *97*, 40–48. [CrossRef]
18. Fu, X.; Dou, C.; Chen, Y.; Chen, X.; Shi, J. Subcellular distribution and chemical forms of cadmium in *Phytplacca americana* L. *J. Hazard. Mater.* **2011**, *186*, 103–107. [CrossRef]
19. Wang, X.; Liu, Y.G.; Zeng, G.M.; Chai, L.Y.; Song, X.C.; Min, Z.Y.; Xiao, X. Subcellular distribution and chemical forms of cadmium in *Bechmeria nivea* (L.) Gaud. *Environ. Exp. Bot.* **2008**, *62*, 389–395. [CrossRef]
20. Qiu, Q.; Wang, Y.; Yang, Z.; Yuan, J. Effects of phosphorus supplied in soil on subcellular distribution and chemical forms in two Chinese flowering cabbage (*Brassica parachinensis* L.) cultivars differing in cadmium accumulation. *Food Chem. Toxicol.* **2011**, *49*, 2260–2267. [CrossRef]
21. Lai, H.Y. Subcellular distribution and chemical forms of cadmium in *Impatiens walleriana* in relation to its phytoextraction potential. *Chemosphere* **2015**, *138*, 370–376. [CrossRef]
22. Garau, G.; Castaldi, C.; Santona, L.; Deiana, P.; Melis, P. Influence of red mud, zeolite and lime on heavy metal immobilization, culturable heterotrophic microbial populations and enzyme activities in a contaminated soil. *Geoderma* **2007**, *142*, 47–57. [CrossRef]
23. Cao, X.; Dermatas, D.; Xu, X.; Shen, G. Immobilization of lead in shooting range soils by means of cement, quicklime, and phosphate amendments. *Environ. Sci. Pollut. Res.* **2008**, *15*, 120–127. [CrossRef] [PubMed]
24. Liu, N.; Jiang, Z.; Li, X.; Liu, H.; Li, N.; Wei, S. Mitigation of rice cadmium (Cd) accumulation by joint application of organic amendments and selenium (Se) in high-Cd-contaminated soils. *Chemosphere* **2020**, *241*, 125106. [CrossRef] [PubMed]
25. Shuman, L. Organic waste amendments effect on zinc fraction of two soils. *J. Environ. Qual.* **1999**, *28*, 1442–1447. [CrossRef]
26. Egene, C.E.; Van Poucke, R.; OK, Y.S.; Meers, E.; Tack, F.M.G. Impact of organic amendments (biochar, compost and peat) on Cd and Zn mobility and solubility in contaminated soil of the Campine region after three years. *Sci. Total Environ.* **2018**, *626*, 195–202. [CrossRef]
27. Ahmad, M.; Rajapaksha, A.U.; Lim, J.E.; Zhang, M.; Bolan, N.; Mohan, D.; Vithanage, M.; Lee, S.S.; Ok, Y.S. Biochar as a sorbent for contaminant management in soil and water: A review. *Chemosphere* **2014**, *99*, 19–33. [CrossRef]

28. Moreno-Jiménez, E.; Esteban, E.; Carpena-Ruiz, R.O.; Lobo, M.C.; Rénalos, J.M. Phytostabilisation with Mediterranean shrubs and liming improved soil quality in a pot experiment with a pyrite mine soil. *J. Hazard. Mater.* **2012**, *201–202*, 52–59. [CrossRef]
29. Xu, C.; Chen, H.X.; Xiang, Q.; Zhu, H.H.; Wang, S.; Zhu, Q.H.; Huang, D.Y.; Zhang, Y.Z. Effect of peanut shell and wheat straw biochar on the availability of Cd and Pb in a soil-rice (*Oryza sativa* L.) system. *Environ. Sci. Pollut. Res.* **2017**, *25*, 1147–1156. [CrossRef]
30. Wang, Y.M.; Tang, D.D.; Zhang, X.H.; Uchimiya, M.; Yuan, X.Y.; Li, M.; Chen, Y.Z. Effects of soil amendments on cadmium transfer along the lettuce-snail food chain: Influence of chemical speciation. *Sci. Total Environ.* **2019**, *649*, 801–807. [CrossRef]
31. Zhang, X.Y.; Zhang, Y.M.; Liu, X.Y.; Zhang, C.Y.; Dong, S.D.; Liu, Q.; Deng, M. Cd uptake by *Phytolacca americana* L. promoted by cornstalk biochar amendments in Cd-contaminated soil. *Int. J. Phytorem.* **2019**, *22*, 251–258. [CrossRef]
32. Chen, Z.S.; Lee, D.Y. Evaluation of remediation techniques on two cadmium-polluted soils in Taiwan. In *Remediation of Soils Contaminated with Metals*; Iskander, I.K., Adriano, D.C., Eds.; Science Reviews: Northwood, UK, 1997; pp. 209–223.
33. Thomas, G.W. Soil pH and soil acidity. In *Methods of Soil Analysis. Part. 3 Chemical Methods*; Sparks, D.L., Page, A.L., Helmke, P.A., Loeppert, R.H., Soltanpour, P.N., Tabatabai, M.A., Johnston, C.T., Sumner, M.E., Eds.; SSSA Inc./ASA Inc.: Madison, WI, USA, 1996; pp. 475–490.
34. Rhoades, J.D. Salinity: Electrical conductivity and total dissolved solids. In *Methods of Soil Analysis. Part 3. Chemical Methods*; Sparks, D.L., Page, A.L., Helmke, P.A., Loeppert, R.H., Soltanpour, P.N., Tabatabai, M.A., Johnston, C.T., Sumner, M.E., Eds.; SSSA Inc./ASA Inc.: Madison, WI, USA, 1996; pp. 417–435.
35. Nelson, D.W.; Sommers, L.E. Total carbon, organic carbon, and organic matter. In *Methods of Soil Analysis. Part 3. Chemical Methods*; Sparks, D.L., Page, A.L., Helmke, P.A., Loeppert, R.H., Soltanpour, P.N., Tabatabai, M.A., Johnston, C.T., Sumner, M.E., Eds.; SSSA Inc./ASA Inc.: Madison, WI, USA, 1996; pp. 961–1010.
36. Sumners, M.E.; Miller, W.P. Cation exchange capacity and exchange coefficients. In *Methods of Soil Analysis. Part 3. Chemical Methods*; Sparks, D.L., Page, A.L., Helmke, P.A., Loeppert, R.H., Soltanpour, P.N., Tabatabai, M.A., Johnston, C.T., Sumner, M.E., Eds.; SSSA Inc./ASA Inc.: Madison, WI, USA, 1996; pp. 1201–1229.
37. Gee, G.W.; Bauder, J.W. Particle-size analysis. In *Methods of Soil Analysis. Part 1. Physical and Mineralogical Method*, 2nd ed.; Klute, A., Ed.; SSSA Inc./ASA Inc.: Madison, WI, USA, 1986; pp. 383–412.
38. EPA/Taiwan. *Method Code No: NIEA S321.65B*; Environmental Protection Administration of Taiwan ROC: Taipei, Taiwan, 2018.
39. Bremner, J.M. Nitrogen-Total. In *Methods of Soil Analysis. Part 3. Chemical Methods*; Sparks, D.L., Page, A.L., Helmke, P.A., Loeppert, R.H., Soltanpour, P.N., Tabatabai, M.A., Johnston, C.T., Sumner, M.E., Eds.; SSSA Inc./ASA Inc.: Madison, WI, USA, 1996; pp. 1085–1121.
40. Kuo, S. Phosphorus. In *Methods of Soil Analysis. Part 3. Chemical Methods*; Sparks, D.L., Page, A.L., Helmke, P.A., Loeppert, R.H., Soltanpour, P.N., Tabatabai, M.A., Johnston, C.T., Sumner, M.E., Eds.; SSSA Inc./ASA Inc.: Madison, WI, USA, 1996; pp. 869–919.
41. Helmke, P.A.; Sparks, D.L. Lithium, Sodium, Potassium, Rubidium, and Cesium. In *Methods of Soil Analysis. Part 3. Chemical Methods*; Sparks, D.L., Page, A.L., Helmke, P.A., Loeppert, R.H., Soltanpour, P.N., Tabatabai, M.A., Johnston, C.T., Sumner, M.E., Eds.; SSSA Inc./ASA Inc.: Madison, WI, USA, 1996; pp. 551–574.
42. De Medici, D.; Komínková, D.; Race, M.; Fabbricino, M.; Součková, L. Evaluation of the potential for caesium transfer from contaminated soil to the food chain as a consequence of uptake by edible vegetables. *Ecotox. Environ. Safe* **2019**, *171*, 558–563. [CrossRef] [PubMed]
43. Yu, T.H.; Hsieh, S.P.; Su, C.M.; Huang, F.J.; Hung, C.C.; Yiin, L.M. Analysis of Leafy Vegetable Nitrate Using a Modified Spectrometric Method. *Int. J. Anal. Chem.* **2018**, *2018*, 1–6. [CrossRef] [PubMed]
44. Baker, D.E.; Amacher, M.C. Nickel, copper, zinc, and cadmium. In *Methods of Soil Analysis. Part 2. Chemical and Microbiological Properties*, 2nd ed.; Page, A.L., Millers, R.H., Keeney, D.R., Eds.; SSSA Inc./ASA Inc.: Madison, WI, USA, 1982; pp. 323–336.
45. Mench, M.J.; Didier, V.L.; Loffler, M.; Gomez, A.; Masson, P. A mimicked in-situ remediation study of metal-contaminated soils with emphasis on cadmium and lead. *J. Environ. Qual.* **1994**, *23*, 58–63. [CrossRef]
46. Ure, A.M. Methods of analysis of heavy metals in soils. In *Heavy Metals in Soils*, 2nd ed.; Alloway, B.J., Ed.; Blackie Academic and Professional: London, UK, 1995; pp. 58–102.

47. Lam, C.M.; Lai, H.Y. Effect of inoculation with arbuscular mycorrhizal fungi and blanching on the bioaccessibility of heavy metals in water spinach (*Ipomoea aquatica* Forsk.). *Ecotox. Environ. Safe* **2018**, *162*, 563–570. [CrossRef]
48. European Commission (EC). Commission Regulation (EU) No 420/2011 of 29 April 2011 amending Regulation (EC) No 1881/2006 setting maximum levels for certain contaminants in foodstuffs. *Off. J. Eur. Union* **2011**, *11*, 3–6.
49. Lai, H.Y.; Chen, B.C. The dynamic growth exhibition and accumulation of cadmium of pak choi grown in contaminated soils. *Int. J. Environ. Res. Public Health* **2013**, *10*, 5284–5298. [CrossRef]
50. Antoniadis, V.; Shaheen, S.M.; Boersch, J.; Frohne, T.; Laing, G.D.; Rinklebe, J. Bioavailability and risk assessment of potentially toxic elements in garden edible vegetables and soils around a highly contaminated former mining area in Germany. *J. Environ. Manag.* **2017**, *186*, 192–200. [CrossRef] [PubMed]
51. European Food Safety Authority. Scientific Opinion of the Panel on Contaminants in the Food Chain on a request from the European Commission of cadmium in food. *ESFA J.* **2009**, *980*, 1–139.
52. Kim, K.; Melough, M.M.; Vance, T.M.; Noh, H.; Koo, S.I.; Chun, O.K. Dietary cadmium intake and sources in the US. *Nutrients* **2019**, *11*, 2. [CrossRef]
53. Lam, C.M.; Chen, K.S.; Lai, H.Y. Chemical forms and health risk of cadmium in water spinach grown in contaminated soil with an increased level of phosphorus. *Int. J. Environ. Res. Public Health* **2019**, *16*, 3322. [CrossRef]

Article

Aqueous Mercury Removal with Carbonaceous and Iron Sulfide Sorbents and Their Applicability as Thin-Layer Caps in Mercury-Contaminated Estuary Sediment

Boon-Lek Ch'ng [1], Che-Jung Hsu [1], Yu Ting [1], Ying-Lin Wang [1], Chi Chen [1], Tien-Chin Chang [2] and Hsing-Cheng Hsi [1,*]

[1] Graduate Institute of Environmental Engineering, National Taiwan University, No. 1, Sec. 4, Roosevelt Rd., Taipei 106, Taiwan; r06541135@ntu.edu.tw (B.-L.C.); zacharyhsu01@gmail.com (C.-J.H.); yuting821216@gmail.com (Y.T.); lynn12783@gmail.com (Y.-L.W.); q2461015@gmail.com (C.C.)

[2] Institute of Environmental Engineering and Management, National Taipei University of Technology, No. 1, Sec. 3, Chung-Hsiao E. Rd., Taipei 106, Taiwan; tcchang@ntut.edu.tw

* Correspondence: hchsi@ntu.edu.tw; Tel.: +886-2-3366-4374

Received: 7 May 2020; Accepted: 12 July 2020; Published: 14 July 2020

Abstract: This study aimed to investigate the Hg removal efficiency of iron sulfide (FeS), sulfurized activated carbon (SAC), and raw activated carbon (AC) sorbents influenced by salinity and dissolved organic matter (DOM), and the effectiveness of these sorbents as thin layer caps on Hg-contaminated sediment remediation via microcosm experiments to decrease the risk of release. In the batch adsorption experiments, FeS showed the greatest Hg^{2+} removal efficiencies, followed by SAC and AC. The effect of salinity levels on FeS was insignificant. In contrast, the Hg^{2+} removal efficiency of AC and SAC increased as increasing the salinity levels. The presence of DOM tended to decrease Hg removal efficiency of sorbents. Microcosm studies also showed that FeS had the greatest Hg sorption in both freshwater and estuary water; furthermore, the methylmercury (MeHg) removal ability of sorbents was greater in the freshwater than that in the estuary water. Notably, for the microcosms without capping, the overlying water MeHg in the estuary microcosm (0.14−1.01 ng/L) was far lesser than that in the freshwater microcosms (2.26−11.35 ng/L). Therefore, Hg compounds in the freshwater may be more bioavailable to microorganisms in methylated phase as compared to those in the estuary water. Overall, FeS showed the best Hg removal efficiency, resistance to salinity, and only slightly affected by DOM in aqueous adsorption experiments. Additionally, in the microcosms, AC showed as the best MeHg adsorber that help inhibiting the release of MeHg into overlying and decreasing the risk to the aqueous system.

Keywords: mercury; methylmercury; salinity; sediment; remediation

1. Introduction

Mercury (Hg) has been known as a global contaminant due to the characteristics of long-range transport in the atmosphere, persistence in the environment, bioaccumulation in the food chain, and adverse effects in human health and ecosystem [1]. The increased accumulation of Hg compounds in sediment may cause the high possibility of Hg being transformed to methylmercury (MeHg) by organisms, which is a neurotoxic compound occurred under anoxic conditions [2–5]. The bioaccumulation and biomagnification of MeHg via food chain transfer may pose a high risk to human through fish consumption [6,7]. Therefore, the strategies for sediment remediation are needed to decrease the Hg contaminant release and the possibility of direct or indirect contact with benthic organisms and water surface.

In-situ capping is a feasible approach to remediate contaminated sediment. The main approach is to allow the sediment left in place but decreasing the chance of further contamination from resuspension of contaminants by the capping layer. This technique could decrease the need for handing sediment and decrease the potential of exposure and consequential spills of sediment. The cost is relatively low as compared to traditional dredging and excavation. Thin layer capping may refer as active capping, which involves a chemically reactive material placing in the subaqueous to sequestrate the emission of contaminants from sediment and decrease the bioavailability, mobility, and toxicity of contaminants [8]. Owing to the physical and chemical sorption properties of reactive materials, the amount of sorbents needed and the thickness of the capping layer to achieve the considerable results are lesser as compared to traditional capping [8]. Besides, remediation technology not only decreases the cost, but also minimizes the exposure of benthic organisms to contaminated sediment and decreases the ecological risk associated with contaminated sediments [9].

The active materials may carry a series of reactions to remove contaminants, adsorption, absorption, and precipitation of contaminants and shift them from the aqueous phase to a solid. It works by increasing the contaminated-solid partition coefficient and extend the isolation time before contaminant breaking through the capping layer. A wide variety of active materials are applied and preferred according to the specific conditions of the remediation site. Carbonaceous materials such as activated carbon (AC), biochars, and surface-modified black carbon could effectively decrease organic contaminants and immobilize Hg. AC has several functional groups such as carboxyl, lactone, and phenolic groups. With its high specific surface area, AC has the potential to be an option for remediation of organic pollutant and Hg-contaminated sites [10–12].

Sulfurized activated carbon (SAC) is generally formed by heating the carbon in the presence of elemental sulfur [13,14] or sulfurous gases [15,16]. SAC provides sulfur-containing functional groups, which show high affinity for Hg compounds to form mercuric sulfide at the surface. Hence, SAC has been verified to further enhance the adsorption capacity of Hg as compared to untreated AC in aqueous adsorption tests [17].

Iron sulfide (FeS) minerals have been widely applied for Hg immobilization on account of the high affinity to Hg ions [18–22]. The mercuric sulfide (HgS) is a stable compound and hardly soluble with a low solubility product constant (Ksp) of 2×10^{-54} for red cinnabar [23] and 4×10^{-54} for black metacinnabar [24,25]. Additionally, FeS can effectively immobilize other divalent metals such as Cd^{2+}, Co^{2+}, Zn^{2+}, and Ni^{2+} through adsorption and coprecipitation [26,27].

The objective of this research attempts to decrease the release of both ionic Hg (Hg^{2+}) and MeHg from sediment to surface water and minimize the negative impacts of Hg contamination on the ecological environment with in-situ thin-layer capping practice, which has been shown to have potentials to decrease Hg contamination in sediment. However, previous application of in-situ capping was primarily focused on systems like contaminated river and lake sediment, there is few studies correlated with wetland and estuary, which are complex systems and effected by tidal flow. Besides, the knowledge of capping material's stability is still limited and easily affected by environmental factors. Therefore, it is necessary to investigate the influence of these environmental factors, such as salinity and dissolved organic material (DOM), on the Hg sorption effectiveness of active capping materials. AC, SAC, and FeS were examined as the capping materials because of their potentially suitable physicochemical properties for Hg sorption. In this study, both aqueous batch sorption experiments and lab-scale vertical up-flow microcosms were conducted to comprehend the impact of environmental factors on the stability of materials and applicability to Hg-contaminated sediment.

2. Materials and Methods

The test sediment was collected from a Hg-contaminated seawater pond in China Petrochemical Development Corporation, An-Shun, Tainan city, Taiwan, designated as An-Shun site. The sediment within 0-15 cm depth was collected using a stainless crab bucket. For the sediment pretreatment procedures, the sediment was air dried in a hood, and the branches and benthic biotas were removed.

The sediment was grounded and sieved through a 20-mesh screen to obtain homogenized sediment. After pretreatment, the sediment was stored at room temperature and covered with a black plastic bag.

Three kinds of sorbents were tested in the experiments. Commercial coconut-shell AC was obtained from Li Jing Viscarb Co. Ltd., Taiwan and sieved to obtain a size range from 18 to 30 mesh. Sieved AC was dried in an oven at 105 °C for 24 h. To obtain SAC, the commercial AC was pretreated with elemental sulfur following the protocol described in Hsi et al. [28]. The prepared SAC had the size range and pretreatment conditions the same as the AC and has been examined in our previous study [29]. FeS was purchased from Sigma-Aldrich.

2.1. *Physicochemical Properties of Sorbents and Sediment*

The physical properties of AC, SAC, and FeS were determined by using a physisorption analyzer (Micromeritics ASAP 2420, Norcross, GA, USA) based on the N_2 adsorption-desorption isotherm at 77K. The Brunauer-Emmett-Teller (BET) equation was used to determine the specific surface area based on ASTM D6556-10 [30], and the micropore surface area and volume were calculated by using the *t*-plot method by using the Jura–Harkins equation: $t = [13.99/(0.034 - \log(p/p_0))]^{0.5}$ [31].

Elemental analyses were conducted to measure the contents of elements including N, C, S, H (Elementar Vario EL cube, Langenselbold, Germany), and O (Thermo Fisher Flash 2000, Waltham, MA, USA) for AC and SAC. The water content of air-dried sediment was measured by the weight method based on the Taiwan Environmental Protection Administration (TEPA) standard method (NIEA S280.62C). Sediment pH value with 1:1 sediment to H_2O ratio was measured by pH meter (Suntex SP-2300, New Taipei City, Taiwan) based on TEPA standard method (NIEA S410.62C). Sediment texture was measured by using the bouyoucos hydrometer method [32]. Sediment organic carbon (OC) content was measured by Walkley-Black wet oxidation [33]. Sediment cation exchange capacity (CEC) was measured by the ammonium acetate method based on the TEPA standard method (NIEA S201.61C). Detailed descriptions pertaining to characterizing the physical and chemical properties of sampled sediment can be found in Supplementary Material.

2.2. *Aqueous Batch Sorption Experiment*

The following steps were conducted in each aqueous batch sorption experiment:

2.2.1. Preparation of Hg Stock Solution

A serial dilution of Hg standard solution was made to the intended concentrations using analyzed reagent-grade Hg standard (1000 μg/mL dilute in nitric acid; Ultra Scientific), and adjust the pH value of the solution to 7.0 for various dosage and Hg concentration experiments. In contrast, the pH value of artificial waters was measured to be 7.6, 8.2, 8.3 for freshwater, estuary water, and seawater conditions, respectively, determined by the ions present (Table S1) to evaluate the effect of salinity and DOM on Hg adsorption. 0.1 M HNO_3 or 0.1 M NaOH were used to adjust the pH of solutions.

2.2.2. Sample Preparation

(1) The Hg stock solution of 50 mL was injected into the glass bottles.
(2) The intended dosage of sorbents was added into the glass bottles and sealed with a rubber plug and aluminum cap.
(3) The bottles were collected, including the samples of triplicate plus the blank.

2.2.3. Adsorption Process

The samples were put into the reciprocating water bath shaker and shaken at 130 rpm at 30 °C for 24 h to achieve an adsorption equilibrium.

In order to determine the suitable dosage of sorbents in the following experiments, the dosages 5, 10, 15, 20, 25, 30, 40 and 50 mg were tested with the Hg concentration of 74 μg/L in 50 mL solution.

To determine the adsorption isotherms of sorbents, 4.3, 25, 56, and 135 μg/L Hg^{2+} were tested with 20 mg of sorbents in 50 mL solution.

In order to study the influence of salinity on Hg removal, the Hg^{2+} solutions (Hg concentration = 197.1 ± 10 μg/L) with three different salinity levels of artificial waters were tested in this study, including freshwater (pH = 7.6 ± 0.1), estuary water (pH = 8.2 ± 0.1), and seawater (pH = 8.3 ± 0.1). The compositions of freshwater and seawater are listed in Table S1. The freshwater was prepared according to Lewis et al. [34]; the seawater was prepared according to Kester et al. [35]. The estuary water was prepared by mixing fresh water and seawater at a 1:1 volume ratio. The concentrations of ion species listed in Table S1 were confirmed by ion chromatography (Metrohm 792 Basic IC, Herisau, Switzerland). Hg(II) species at various salinity levels were simulated by using MINTEQ 3.1 and listed in Table S2.

To evaluate the effect of DOM on sorbents' Hg removal efficiency, the DOM solution was prepared by humic acid (HA, Sigma Aldrich, Saint Louis, MO, USA). HA of 10 mg was dissolved in ultrapure water and pre-adjusted to pH 7 by 0.1 M NaOH. The solution was stirred for 1 h to promote fast dissolution, and then filtered through a 0.2 μm mixed cellulose ester filter (DISMIC-25AS, Toyo Roshi Kaisha, Tokyo, Japan). The filtrate was collected and stored at 4 °C refrigerator before using. The actual concentration of DOM was verified by a Total OC analyzer (OI Analytical Aurora 1030W, College Station, TX, USA). The DOM was controlled at a concentration of approximately 2.5 mg-C/L and the Hg concentration is 196.2 ± 5 μg/L.

2.2.4. Preservation of Sample and Analysis

(1) Each sample was preserved with 0.5% $BrCl_2$ and estimated by cold vapor atomic fluorescence spectroscopy (CVAFS; Brooks Rand Automated Total Mercury System, Seattle, WA, USA).

(2) The aqueous Hg removal efficiency by sorbents was calculated by the following Equation (1):

$$R = \frac{C_0 - C_t}{C_0} \times 100\%, \quad (1)$$

where R (%) is the Hg removal efficiency of sorbents, C_0 (ng/L) is the initial Hg concentrations detected in blank solution, and C_t (ng/L) is the concentration of remaining Hg at any time.

Equation (2) was used to determine the partitioning coefficient (K_D) for Hg adsorption by the adsorbent:

$$K_D = \frac{q_e}{C_e} \quad (2)$$

where q_e is the equilibrium Hg adsorption capacity (mg/g) and C_e is the equilibrium Hg concentration (μM). The calculated values of K_D were listed in Table S3.

2.3. *Laboratory Microcosm Experiments*

The microcosm was established according to Ting et al. [17] with modifications. Vertical up-flow columns were used to stimulate the release of Hg compounds and examine the efficiency of sorbent cappings on Hg sequestration (Figure 1). The dimensions of the column with an internal height of 15 cm and an internal diameter of 6 cm with glass fiber at the bottom. To investigate how the Hg-contaminated estuary sediment affected by salinity and to understand the stabilizing efficiency of AC, SAC, and FeS cappings, two systems, the freshwater and estuary systems, were set up. Each of the four columns contained 300 g of dried Hg-contaminated sediment from the An-Shun site. The total number of column is 8. Column A was capped with AC (9 g; i.e., 3 wt% AC added); column B was capped with SAC (9 g; i.e., 3 wt% SAC added); column C was capped with FeS (9 g; i.e., 3 wt% FeS added); column D was without capping as the control unit.

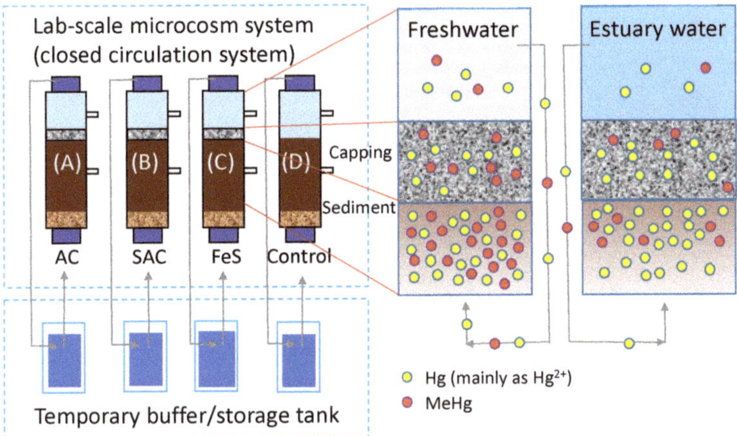

Figure 1. The diagram of vertical up-flow microcosms construction.

To start up the microcosms, dried Hg-contaminated sediment was firstly added into the column, filled with waters and waited for 24 h to settle. The microcosms were activated to circulate the waters in the column for sediment acclimation and stabilization, and counted as the operation day 1. The capping materials were applied on day 25. The total volume of water used for each experiment was 1.5 L. The water first entering into the bottom of the column was referred as inflow, then vertically moved upwards to fill the column, and then discharged through the outlet. The effluent water were stored in a temporary buffer tank and recirculated into the bottom of column in a closed system by using a peristaltic pump (Lead Fluid BT100S, Hebei, China) to maintain a flow rate of 1 mL/min. The reason that the effluent water was recirculated back to the microcosm column is to establish the mass balance of THg partitioning in various phases (i.e., sediment, capping material, overlying water) in a closed system and evaluate the THg accumulation ability of the various capping materials.

Periodic sampling was in progress while the microcosms were in operation. Each time, 100 mL of water sample was collected from the tube of outflow. After that, the temperature, pH (SunTex SR-2300, New Taipei City, Taiwan), dissolved oxygen (DO; Extech EXStik DO600, Nashua, NH, USA), electrical conductivity (EC; Taina EZDO 6021, Taichung, Taiwan), oxidation reduction potential (ORP; SunTex SR-2300, New Taipei City, Taiwan), trace metals concentration, total Hg (THg), MeHg, DOM, and anions were analyzed. The temporary buffer tank was then refilled with the artificial waters to maintain a constant water amount.

To measure the trace compound concentration, including metals, THg, MeHg, DOM, and anions, the water samples were filtered with 0.45 μm mixed cellulose ester filter (DISMIC-25AS, Toyo Roshi Kaisha, Tokyo, Japan). Trace metals were determined by Inductively couple plasma optical emission spectrometry (Agilent 700, Santa Clara, CA, USA) after the sample was acidified with 0.15% HNO_3. Water samples for THg analysis were preserved by adding 0.5% $BrCl_2$ solution and stored in 20 mL glass bottles. THg in sediment and water was analyzed following the USEPA Method 1631 and NIEA W331.50B protocols by using CVAFS. The water samples were preserved by adding 0.2% HCl and stored in 20 mL amber glass bottles in MeHg analysis. MeHg in sediment and water was analyzed following the USEPA Method 1630, NIEA W341.60B and NIEA W540.50B procedures by using gas chromatography/CVAFS (Brooks Rand MERX Integrated Automated MeHg Analyzer, Seattle, WA, USA).

Recovery of Hg in the microcosm system was evaluated based on the Hg concentration in overlying water, cap materials, and sediment. To determine the Hg concentration in cap materials and sediment, digestion was first conducted in a microwave system (Ethos 1600, Milestone, Shelton, CT, USA) with a power setting of 800 W (USEPA method 3051a). After digestion, the THg (μg/mg) in the solid phase was determined using CVAFS (USEPA method 245.7). QA/QC of data were confirmed based on Hsu et al. [29] and Wang et al. [36]. The recovery values for the QC samples of sediment of THg (NIST 2709a, 0.9 mg/kg) and MeHg (SQC-1238, 10 μg/kg) were 98.6 and 90.0%, respectively. For spiked sediment, the recovery values of THg and MeHg were 94.8 and 95.2%, and were 107.1 and 101.4% for water, respectively. The coefficient of determination (R^2) of CVAFS for the aqueous Hg was regularly kept larger than 0.998, the recovery was within 96.2–120%, the precision was within 0.04–5.60%.

2.4. Statistical Analysis

A one-way ANOVA, followed by a least significant difference (LSD) test ($p < 0.05$), was used to determine the significance differences among microcosm tests with various capping materials (IBM SPSS statistics).

3. Results and Discussion

3.1. Physicochemical Properties of Sorbents

The physical and chemical properties of the three test sorbents, AC, SAC, and FeS, were summarized in Table 1. Because FeS is a non-porous mineral, its surface area and pore volume were much smaller than those of AC and SAC. Elemental analyses of AC and SAC showed that the oxygen and sulfur contents in SAC were increased after impregnation of sulfur on AC, hence SAC should be more favored for Hg^{2+} uptake because sulfur-containing groups on SAC had high affinity towards mercuric sulfide formation. The oxygenated groups are also known beneficial for Hg adsorption. The increase in the oxygen content after sulfur impregnation could be due to the reaction of S with C to form vaporized CS_2 evolved from the SAC surface.

Table 1. The physicochemical properties of sorbents.

	S_{BET} (m^2/g) *	S_{Micro} (m^2/g) *	V_{Total} (cm^3/g) *		V_{micro} (cm^3/g) *
AC	1024.1	634.1	0.540		0.284
SAC	903.3	528.9	0.502		0.267
FeS	2.811	0.356	0.04		-
	C (%)	**H (%)**	**O (%)**	**N (%)**	**S (%)**
AC	78.3	1.61	7.72	0.791	0.672
SAC	74.9	1.80	13.8	0.36	5.75
FeS	-	-	-	-	36.4

* S_{BET}: specific BET surface area; S_{Micro}: micropore area; V_{Total}: total pore volume; V_{micro}: micropore volume.

3.2. Aqueous Batch Adsorption Experiment

3.2.1. The Effects of the Sorbent Dosage

The influence of sorbent dosage on Hg removal has been investigated to find out the appropriate dosage of sorbents as a basis for subsequently study. The sorbent dosage determines the sorption capacity of sorbents for a given concentration of Hg^{2+} because it controls the sorbent-sorbate equilibrium of a system [37]. The effects of sorbents dosage on Hg sorption were studied in the dosage range of 5–60 mg in 50 mL Hg^{2+} solution. The results of AC, SAC, and FeS dosage are presented in Figure 2, and all experimental data are performed in triplicate. The Hg sorption capacity of AC first increased and then decreased as the Hg removal efficiency increased, due to the increment of AC dosage would provide more sorption active sites to take up Hg in a fixed Hg^{2+} initial concentration. The SAC and FeS showed a similar tendency as AC. Based on the result, the suggested optimum dosages for AC, SAC and FeS at a fixed initial concentration of Hg^{2+} were 30, 10, and 10 mg, respectively. Owning to Hg removal efficiency may be affected by both the properties and dosage of sorbents, the normalization and comparison of different types sorbents in the same mass benchmark were needed. Therefore, the appropriate dosage of sorbents was determined as 20 mg for the subsequent experiments in this research.

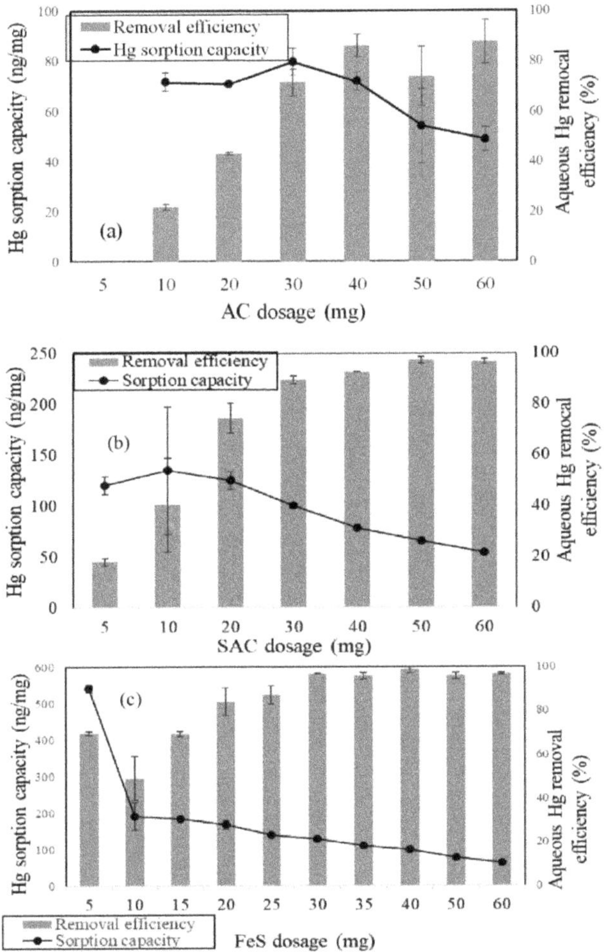

Figure 2. Effect of (**a**) AC; (**b**) SAC; (**c**) FeS dosage on Hg removal efficiency.

3.2.2. Effect of Initial Concentration of Hg^{2+}

The batch experiments were tested within the initial concentration of Hg^{2+} from 5 to 135 μg/L using a diluted standard Hg^{2+} solution. The Hg sorption capacity of sorbents and the initial Hg^{2+} concentration is shown in Figure 3. The experimental results showed that the Hg sorption capacity of sorbents increased with the increment of the initial concentration of Hg^{2+}, with linear adsorption behaviors within the test concentration range. The FeS could maintain a high Hg removal efficiency of up to 90% in the range of the initial concentration of Hg^{2+} given, which illustrates that FeS is an excellent sorbent for Hg removal followed by SAC and AC.

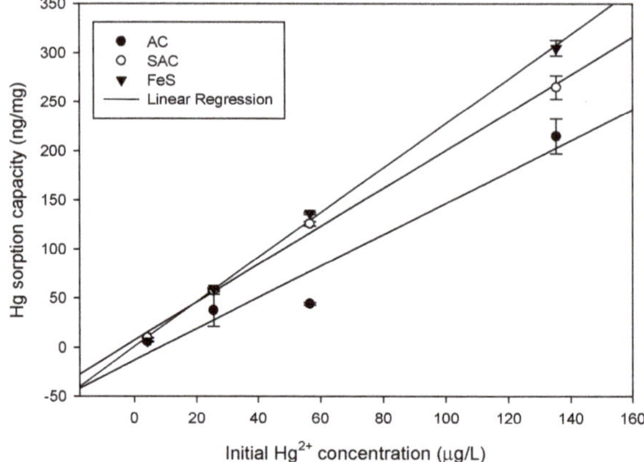

Figure 3. Hg sorption at different Hg^{2+} concentrations.

3.2.3. Effect of Salinity

The salinity level of water in the Hg-contaminated site may affect the Hg removal ability of sorbents. Thereby, three artificial synthetic waters, including freshwater, estuary water, and seawater with different salinity levels, were prepared to study the effects of salinity on Hg removal ability of sorbents. The Hg removal efficiency of sorbents for each water system are presented in Figure 4, indicating that FeS had the largest Hg removal efficiency, followed by SAC and then AC. From the lowest salinity (freshwater) to the highest salinity level (seawater), the Hg removal efficiency increased for both AC and SAC. Although the effect of salinity levels on FeS was insignificant, the Hg removal efficiency of FeS was still the highest as compared to AC and SAC. The calculated K_D values are listed in Table S3, which shows that K_D values increased as the salinity increased. FeS also performed the largest K_D values, followed by SAC and then AC.

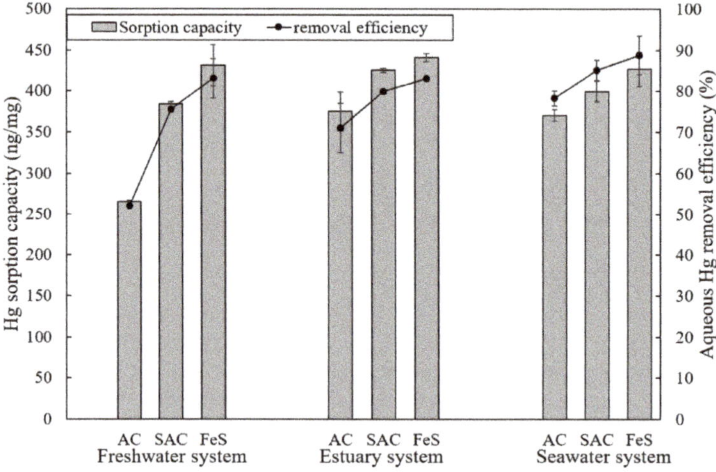

Figure 4. Comparison of various salinity levels affecting the sorbents' Hg removal.

Notably, our results were contrary to those in previous studies, which have shown that a high level of salinity may decrease the removal efficiency of Hg by sorbents [38,39]. The previous studies have demonstrated that an increase in NaCl concentration would decrease the sorption efficiency of

AC and kaolin. However in our research, the different salinity levels were prepared by adding various salts to form the artificial waters. Hence, the complexation and species of Hg in this study may be more complicated than the cases with only presence of NaCl. Table S2 displays the simulated speciation of Hg(II) compounds, indicating that the fraction of $Hg(OH)_2$ in the freshwater system (i.e., 95.97%) was significantly higher than that in both systems of estuary water (i.e., 0.094%) and seawater (i.e., 0.014%). $Hg(OH)_2$ has been reported that it could easily decompose to the elemental form as Hg^0, Hg^0 is more difficult to be captured from the water solution than other oxidized form because of its extremely low solubility (i.e., 5.6×10^{-5} g/L) [29]. Additionally, Thiem et al. [40] showed that the addition of calcium ion would enhance Hg removal of AC. They speculated that the calcium ion may react with the surface group on AC to form a new adsorption site, leading to an increment of Hg removal capacity in the solution. Besides, according to the K_D values, the increase of salinity has a positive effect on the partition behavior of aqueous Hg to the adsorbent; furthermore, FeS has a fabulous capability for converting Hg from the liquid phase to the solid phase.

3.2.4. Effect of Dissolved Organic Matter on Hg Sorption at Various Salinity Levels

DOM is widely spread in environments and may control a number of essential processes relevant for Hg cycling. In this study, artificial waters with a DOM concentration of 2.6 mg-C/L was prepared with three salinity levels, and the results are shown in Figure 5. For the addition of DOM to the salinity test, the Hg removal by sorbents was relatively decreased as compared with those in the salinity test (Figure 4). In freshwater system, the sorption capacity of AC, SAC, and FeS was decreased from 286 to 137 ng/mg, 384 to 270 ng/mg, and 431 to 287 ng/mg, respectively. As for the estuary system, the sorption capacity of AC, SAC and FeS was decreased from 401 to 286 ng/mg, 408 to 366 ng/mg, and 441 to 403 ng/mg, respectively. While the sorption capacity of AC, SAC, and FeS were decreased from 355 to 302 ng/mg, 400 to 322 ng/mg, and 427 to 363 ng/mg in the seawater system, respectively (Figure 6). In Table S3, K_D values also decreased significantly with the presence of DOM. Therefore, DOM may inhibit the Hg adsorption by complexation mechanism because it can compete with sorbents and complex with Hg^{2+} [41–44]. Hg^{2+} may form complex with organic thiol groups in DOM [45,46]. The phenolic hydroxyl groups in DOM may also complex with Hg easily to form a stable chelate that restrained the Hg adsorption [47].

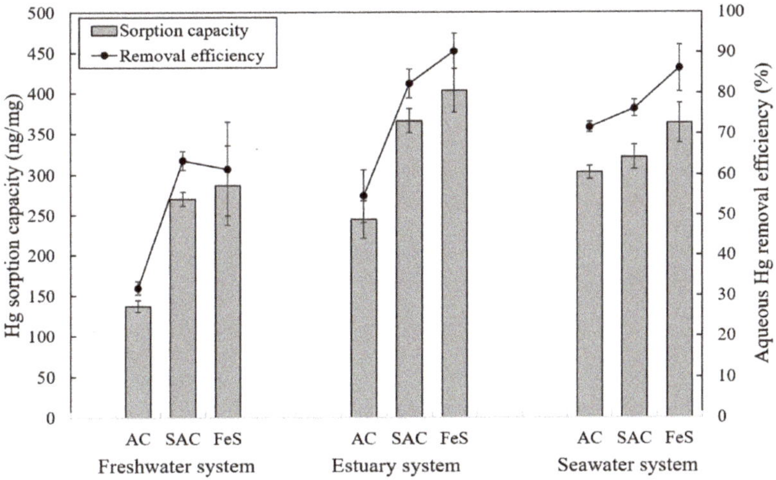

Figure 5. Comparison of the Hg sorption capacity at various salinity levels affected by DOM.

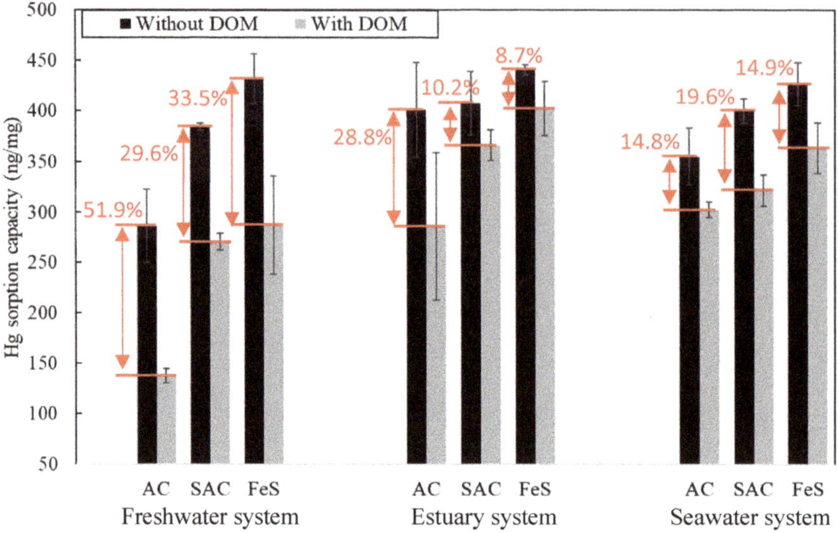

Figure 6. Comparison of the influence of various salinity levels and DOM on Hg sorption capacity in the aqueous batch experiments.

3.3. Laboratory Microcosm Experiments

3.3.1. Physicochemical Properties of Sediments Used in Microcosm Experiments

The physicochemical properties of the An-Shun site sediment are summarized in Table 2. The pH value of the sediment was about 7.5, slightly alkaline due to the estuary water system of that region. The slightly alkaline condition may favor to immobilize and decrease the toxicity of most metals as they may precipitate in the alkaline environment. The OC content in the An-Shun site sediment was about 0.8 wt%, which was relatively low when compared to other sediments. The organic matter in the sediment may potentially affect the interaction between pollutants and sediment. The sand, silt and clay contents of the sediment were 71.4, 14.3, and 14.3 wt%, respectively. The An-Shun site sediment could thus be classified as a sandy loam according to the soil texture classification of the United States Department of Agriculture (USDA) (Figure S2). In general, the texture of sediment is tightly correlated with OC content and CEC. The fine-textured sediment may contain high OC and CEC due to its large surface areas, whereas the OC and CEC of sandy-textured sediment was relatively low. The results of physicochemical properties analysis of An-Shun site sediment were consistent with the fact that a sandy loam showed low OC content and low CEC.

Table 2. The physicochemical properties of An-Shun site sediment.

Parameters	Value
Water content (wt%)	2.33 ± 0.01
pH	7.5 ± 0.03
Organic carbon (wt%)	0.80 ± 0.65
Cation exchange capacity (cmol(+)/kg)	3.3 ± 0.2
Sand (wt%)	71.4 ± 2.9
Silt (wt%)	14.3 ± 0.0
Clay (wt%)	14.3 ± 2.9
THg (mg/kg)	76.0 ± 2.59
MeHg (μg/kg)	1.17 ± 0.31

3.3.2. Environmental Factor Variation during Microcosms Operation

The vertical up-flow microcosms with a closed circulation system are set up as shown in Figure 1 and Figure S1. Two kinds of waters were applied in the microcosms, including freshwater and estuary water. The microcosms were operated for 65 days. The operation time from days 1 to 25 was the process of sediment acclimation. After that, the capping materials were delivered to the microcosms and the water was collected periodically from the outflow as the overlying water sample. The appearance of the microcosms, which was capped with the AC, SAC, or no cap were seen to be clear, while the column capped with FeS was found to gradually appear in brown inside the microcosms. It has been speculated that the addition of FeS may promote a chemical reaction to form iron hydroxide or iron oxide, which was found attached on the wall of the column and the buffer tank.

The recorded temperature in the microcosms is shown in Figure S3. The range of temperature was between 20 and 27 °C during the test period between winter and spring. The DO results are shown in Figure S4. The DO value at the beginning of the experiment was approximately 4 mg/L, started to decrease and remained stable at 3 ± 0.5 mg/L as the operation time extended. This result could be explained by the microbial activity in the microcosms. The microbes would consume DO to carry out respiration process, thus its concentration decreased at the beginning. Owning to the operational defects, the oxygen outside the microcosm fluxed in and caused the increment of DO. The pH value of overlying water is shown in Figure S5. It was between 7.6 and 7.9 in all microcosms.

The results of EC in the freshwater system and estuary system are shown in Figure S6. In the freshwater system, the EC of overlying waters of microcosms was around 2300–3270 µS/cm. The EC values decreased as the operation time increased because the microcosms would be refilled with the artificial freshwater and get diluted after sampling periodically (Figure S6a). Moreover, the An-Shun site sediment was estuary sediment, thus it contained a portion of salt and contributed to the background levels of salinity. In the estuary water system of microcosms, the results of EC are shown in Figure S6b. The EC ranged from 37,000 to 38,000 µS/cm.

The ORP results of the overlying water from freshwater system and estuary system are shown in Figure S7. In the freshwater system, the ORP of overlying water of microcosms was around 40–160 mV. As for the estuary system, the ORP of overlying water of microcosms was about −22–145 mV.

The results of DOM in the overlying water are shown in Figure S8. The DOM content in the overlying water was low, mainly related to the low DOM content in An-Shun site sediment. The An-Shun site sediment contained less OC content according to the previous results shown; consequently, less DOM would be released from the sediment to the overlying water.

The total Fe in overlying water is shown in Figure S9. In the freshwater system, the microcosm capped with FeS would contribute significant dissolved Fe to the overlying water while the concentration of Fe in the microcosm with no capped was also high. The Fe content of those treatment were higher if compared to microcosms with AC and SAC. The similar trend was also observed in the estuary system.

3.3.3. Sequestration of Aqueous THg and MeHg by Thin Layer Capping

The effectiveness of capping materials to immobilize Hg is shown in Figure 7 and Table S4. For the freshwater system, the THg immobilization abilities of three capping materials greatly varied as compared to that of the no-capped control unit. The THg reduction efficiencies of the AC, SAC, and FeS, which were calculated based on the initial THg concentration of the microcosm started and the THg concentration in a given sampling date, fluctuated between −22–82%, −30–78% and −87–62%, respectively, in comparison to the control unit. As for the estuary system, THg reduction efficiencies of the AC, SAC and FeS reached −33–49%, 28–60% and −24–44%, respectively, in comparison to that of the control unit. Therefore, the fluctuation and uncertainty for Hg removal of capping sorbents were observed in both freshwater and estuary systems. According to our previous studies [17,48], the recommended amendment dosage of sorbents for Hg-sediment remediation was approximately 1–5%. The dosage of sorbents being used in this study was 3%, which was supposedly to show significant effect on Hg removal, but the results seemed not be similar to our previous studies. The horizontal flow microcosm set up was mostly used to stimulate the emission of Hg from the sediment according to the previous studied [48]. However, a different design, vertical up-flow microcosm was adapted in this study, in which the water moved in an up-flow direction and then recirculated back into the columns in a closed system. Therefore, a dynamic equilibrium among the sediment, capping material, and overlying water achieved in the closed circulation system. The Hg in the sediments will be released to the water continuously until an equilibrium was reached. The same process happened in the interface between water and capping material. These processes will lead to the depletion of Hg on the sediments because it was diffused and transferred to the capping materials. Owning to these process, the reduction of Hg in the overlying water of the microcosm was less evident, which was also demonstrated by the ANOVA analysis (Table S4); instant and marked releases of Hg from the caps were occasionally found. The dosage of sorbents applied in this microcosms may need to be reconsidered as it might not be capable to sorb Hg efficiently due to the characteristic of this microcosms system.

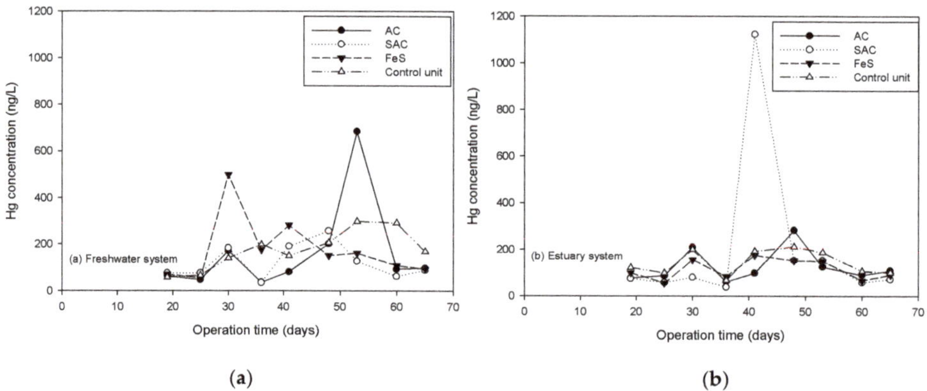

Figure 7. THg concentration in overlying water of (**a**) freshwater system and (**b**) estuary system.

The Hg concentrations in sediment, sorbents and overlying water after 65 days in the microcosms are shown in Table 3. The Hg concentrations in the freshwater sediments at the end of experiment for AC, SAC, FeS and control unit were 65.35 ± 0.65, 68.87 ± 0.46, 50.38 ± 1.03 and 68.77 ± 2.03 mg/kg, respectively. In the estuary system, the Hg concentrations of sediment for AC, SAC, FeS and control unit were 56.16 ± 2.87, 57.55 ± 0.39, 54.80 ± 2.25 and 69.20 ± 4.02 mg/kg, respectively. The initial Hg content in sediment of the An-Shun site was 76.00 ± 2.59 mg/kg, thus, there were varying losses of Hg from sediment with different capping amendment in both freshwater and estuary systems. The sediment can be a reservoir of Hg, which would provide Hg to the water, capping materials, and the microcosm's auxiliary equipment until an equilibrium reached. Hence, the decrease of Hg in the sediment of the study is reasonable. In the freshwater system, the loss of Hg in microcosms capped with AC and SAC was similar to that in the control unit while the loss of Hg in FeS capping was slightly higher than that in the control unit, AC, and SAC capping. As for the estuary system, the Hg loss from sediment with capping materials was decreased significantly as compared to that in the control unit. On the other hand, the contents of Hg in the sorbents of AC, SAC, and FeS were 4.84 ± 0.81, 6.86 ± 3.52, and 37.66 ± 8.34 mg/kg, respectively, in freshwater system. In the estuary system, the contents of Hg in the sorbents of AC, SAC, and FeS were 3.88 ± 1.82, 2.89 ± 1.49, and 43.83 ± 13.04 mg/kg, respectively. The Hg content in FeS was significantly greater as compared to those in AC and SAC in both freshwater and estuary systems. Notably, the Hg sorption capacity of sorbents in the batch experiments was relatively high when compared to those for the sorbents applied in the microcosms. Although the Hg concentration of overlying water, approximately 200 ng/L was relatively low as compared to the batch experiment test, the mechanism of Hg sorption might be different at the low Hg concentration. The recovery of Hg in the microcosms was approximately 67–90% in both freshwater and estuary systems. The low recovery in some cases can be explained by the phenomenon of bacteria respiration, which could produce a large portion of batting biomass and colloids that cause Hg to attach onto. Moreover, the Hg might be adhered to the surface of the microcosm's auxiliary equipment, such as column surface, buffer tank, pipelines, and others that may cause the decreasing recovery of Hg (Figure S1). Therefore, although FeS showed great Hg sorption capacity during the batch and caused Hg partitioning from sediment to the cap during microcosm test, Hg may be released from cap again, through redissolution or release of small-scale FeS particles containing Hg, leading to not only the low recovery but also the risk of unexpected release of Hg for sediment with vertical up-flow.

Table 3. Recovery calculation based on the Hg content in sediment, sorbents and overlying water after 65 days in the microcosms.

	Sediment [1]		Sorbents		Overlying Water	Recovery [2]
	mg/kg	mg	mg/kg	mg	mg	%
F-AC	65.35 ± 0.65	19.61	4.84 ± 0.81	0.04	0.002	86.19
F-SAC	68.87 ± 0.46	20.66	6.86 ± 3.52	0.06	0.02	90.89
F-FeS	50.38 ± 1.03	15.11	37.66 ± 8.34	0.34	0.02	67.78
F-Control	68.77 ± 2.03	20.63	-	-	0.02	90.50
E-AC	56.16 ± 2.87	16.85	3.88 ± 1.82	0.03	0.002	74.05
E-SAC	57.55 ± 0.39	17.27	2.89 ± 1.49	0.03	0.003	75.85
E-FeS	54.80 ± 2.25	16.44	43.83 ± 13.04	0.39	0.002	73.84
E-Control	69.20 ± 4.02	20.70	-	-	0.002	90.80

[1] The Hg content in sediment for the day 65. [2] Recovery was obtained by the sum of Hg content in sediment, sorbents and overlying water divided by the total Hg content in initial sediment (i.e., 76.0 ± 2.59 mg/kg and 22.8 mg) then multiplying by 100.

The results of MeHg content in the overlying water of freshwater system and estuary system are shown in Figure 8. In the freshwater system, the amounts of MeHg in the overlying water from microcosms capped with AC, SAC, and FeS were significantly decreased when compared to that of the control unit; the MeHg concentrations of AC, SAC, FeS, and control unit were 0.04–0.70, 0.14–0.94, 0.13–1.64, and 2.26–11.35 ng/L, respectively. In contrast, the MeHg concentrations of AC, SAC, FeS, and control unit were 0.10–0.14, 0.15–0.87, 0.04–2.77, and 0.14–1.01 ng/L in the estuary system, respectively. Therefore, the production of MeHg under the treatment of AC and SAC capping was decreased tremendously, while the reduction efficiency of MeHg in the microcosms capped with FeS was not significant.

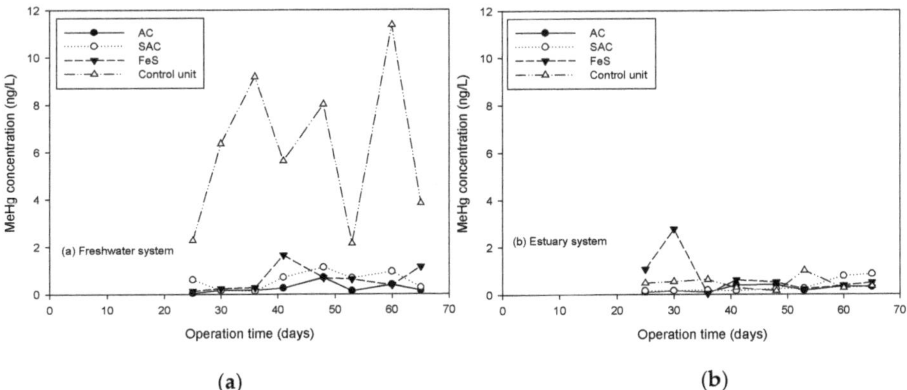

(**a**) (**b**)

Figure 8. MeHg concentration in overlying water of (**a**) freshwater system and (**b**) estuary system.

The production of MeHg of the control unit in the freshwater system was significant as compared with that in the estuary system. A previous study showed that the salinity level of the environment was negatively correlated with MeHg production [49]. In the high salinity condition, Hg has tendency to form complex with chlorine salt and sulfurous compounds, reducing the trends of Hg methylation [50]. The researchers suggested that a high-salinity condition may inhibit Hg methylation because sulfate with high concentration would reduce to toxic sulfides by microorganisms, and poisons the Hg methylating bacteria, reducing the formation of MeHg as well. In the freshwater system, the MeHg reduction efficiency in microcosms capped with AC was the best, SAC followed, and FeS the least, suggesting that the SAC and FeS provided a portion of the iron and sulfur elements to dissolve into water and enhanced the methylation ability of sulfate-reducing bacterial and Fe(III)-reducing bacteria. As a result, the MeHg reduction efficiency in microcosms capped with SAC and FeS was poorer as compared to that with AC.

The MeHg contents in the sediment are shown in Table 4. The formation of MeHg in the control unit of freshwater system was higher than that in the estuary system. The formation of MeHg capped with AC and FeS was low when compared to that in the control unit in freshwater system, except for the microcosm capped with SAC, which was slightly higher. As a results, the MeHg released to overlying water in freshwater system was inhibited by both limiting the MeHg formation and the effectiveness of caps. In the estuary system, the MeHg contents in both capped with sorbents and control unit were similar and smaller than those in freshwater system.

Table 4. The MeHg content in sediment on day 65.

Freshwater System	MeHg (μg/kg)	Estuary System	MeHg (μg/kg)
AC	1.43 ± 0.75	AC	1.03 ± 0.52
SAC	5.42 ± 1.92	SAC	1.16 ± 0.07
FeS	2.61 ± 0.65	FeS	2.07 ± 0.52
Control unit	3.97 ± 1.68	Control unit	1.99 ± 1.15

4. Conclusions

In this study, the aqueous batch experiments with the amendment of AC, SAC, and FeS sorbents were first carried out to comprehend the Hg removal efficiency in Hg-contaminated sediments influenced by salinity and DOM. The microcosms were then set up to examine the performance of these capping sorbents on Hg-contaminated sediment remediation. The experimental results showed that FeS on Hg removal was not significantly affected by salinity levels and maintained with high removal efficiency. The Hg removal efficiency of AC and SAC increased as salinity increased. In contrast, the Hg removal efficiency of sorbents decreased with the addition of DOM at different salinity levels because DOM competed with sorbents and may occupy the adsorption site, thus inhibited the Hg uptake by sorbents. The microcosm experiments showed that the THg immobilization abilities of three capping sorbents greatly varied as compared to that of control unit. The MeHg concentration of overlying water in the freshwater microcosm with no cap was higher than that in the estuary system. Therefore, Hg compounds in the freshwater system may be more bioavailable to microorganisms in methylated phase as compared to those in the estuary system. To summarize, the capping materials including AC, SAC, and FeS effectively decreased the concentration of overlying water MeHg in the freshwater system of microcosms. Because the production of MeHg in estuary system was low, the efficiency of materials on MeHg sorption was insignificant.

We suggest that future studies should be focused on scale-up design using large microcosms. Notably, because FeS showed the best Hg removal efficiency, resistance to salinity, and only slightly affected by DOM in aqueous adsorption experiments and AC showed as the best MeHg adsorption material, a "mixing cap" using both FeS and AC should be examined and the optimal mixing ratio should be obtained. A mixing cap of FeS and AC may also help preventing the leaching out of FeS from the cap layer, which was observed in our microcosm study.

It is also worth noting that for the vertical up-flow system, the accumulated Hg in capped layer may eventually breakthrough, which could cause sudden concentration shock that leads to instant risk of exposure. Long-term microcosm operation is critical and should be further conducted to obtain design parameters on subsequent pilot tests or full-scale application.

Supplementary Materials: The following are available online at http://www.mdpi.com/2073-4441/12/7/1991/s1, Table S1. Freshwater and seawater ion species concentration, Table S2. Hg(II) speciation at various salinity levels, Table S3. The partitioning coefficients for Hg adsorptions at various salinity levels, Table S4. A one-way ANOVA or one-way ANOVA on ranks based on normality test, followed by a post hoc test ($p < 0.05$) used to determine the significance differences among various sorbents, Figure S1. Photos of the microcosms on (a) day 25 as the capping materials were initially applied and (b) on day 65, Figure S2. An-Shun site sediment texture, Figure S3. The temperature of microcosms (the symbol F refers to freshwater system and the symbol E refers to estuary system), Figure S4. The dissolved oxygen for the microcosms (the symbol F refers to freshwater system and the symbol E refers to estuary system), Figure S5. The pH value variation of microcosms (the symbol F refers to freshwater system and the symbol E refers to estuary system), Figure S6. The electricity conductivity of overlying water in (a) freshwater system and (b) estuary system, Figure S7. The oxidation reduction potential of overlying water in (a) freshwater system and (b) estuary system, Figure S8. DOM variation of microcosms for (a) freshwater system; (b) estuary system, Figure S9. The total Fe variation of microcosms in (a) freshwater system and (b) estuary system.

Author Contributions: Conceptualization, B.-L.C., Y.T., T.C.C., and H.-C.H.; methodology, B.-L.C., C.-J.H., Y.T., Y.-L.W., and H.-C.H.; formal analysis, B.-L.C. and C.C.; resources, H.-C.H.; writing—original draft preparation, B.-L.C. and C.-J.H.; writing—review and editing, Y.-L.W. and H.-C.H.; supervision, T.-C.C. and H.-C.H.; funding acquisition, H.-C.H. All authors have read and agreed to the published version of the manuscript.

Funding: This research and the APC were funded by the Environmental Protection Administration, Taiwan under Grant no. 08BT547001 and the Ministry of Science and Technology of Taiwan, Taiwan under Grant no. MOST 105-2221-E-002-008-MY3.

Acknowledgments: We greatly appreciate the financial and technical supports from the Environmental Protection Administration, Taiwan and the Ministry of Science and Technology of Taiwan, Taiwan. The opinions expressed in this paper are not necessarily those of the sponsors.

Conflicts of Interest: The authors declare no conflict of interest.

References

1. Nriagu, J.O. *Biogeochemistry of Mercury in the Environment*; Elsevier/North-Holland Biomedical Press: Amsterdam, The Netherlands, 1979.
2. Compeau, G.C.; Bartha, R. Sulfate-reducing bacteria: Principal methylators of mercury in anoxic estuarine sediment. *Appl. Environ. Microbiol.* **1985**, *50*, 498–502. [CrossRef]
3. Gilmour, C.C.; Henry, E.A.; Mitchell, R. Sulfate stimulation of mercury methylation in freshwater sediments. *Environ. Sci. Technol.* **1992**, *26*, 2281–2287. [CrossRef]
4. Gilmour, C.C.; Podar, M.; Bullock, A.L.; Graham, A.M.; Brown, S.D.; Somenahally, A.C.; Johs, A.; Hurt, R.A.; Bailey, K.L.; Elias, D.A. Mercury methylation by novel microorganisms from new environments. *Environ. Sci. Technol.* **2013**, *47*, 11810–11820. [CrossRef] [PubMed]
5. Podar, M.; Gilmour, C.C.; Brandt, C.C.; Soren, A.; Brown, S.D.; Crable, B.R.; Palumbo, A.V.; Somenahally, A.C.; Elias, D.A. Global prevalence and distribution of genes and microorganisms involved in mercury methylation. *Sci. Adv.* **2015**, *1*, e1500675. [CrossRef] [PubMed]
6. Fitzgerald, W.F.; Lamborg, C.H.; Hammerschmidt, C.R. Marine biogeochemical cycling of mercury. *Chem. Rev.* **2007**, *107*, 641–662. [CrossRef] [PubMed]
7. Stein, E.D.; Cohen, Y.; Winer, A.M. Environmental distribution and transformation of mercury compounds. *Crit. Rev. Environ. Sci. Technol.* **1996**, *26*, 1–43. [CrossRef]
8. Zhang, C.; Zhu, M.Y.; Zeng, G.M.; Yu, Z.G.; Cui, F.; Yang, Z.Z.; Shen, L.Q. Active capping technology: A new environmental remediation of contaminated sediment. *Environ. Sci. Pollut. Res.* **2016**, *23*, 4370–4386. [CrossRef] [PubMed]
9. McDonough, K.M.; Murphy, P.; Olsta, J.; Zhu, Y.; Reible, D.; Lowry, G.V. Development and placement of a sorbent-amended thin layer sediment cap in the Anacostia River. *Soil Sediment Contam.* **2007**, *16*, 313–322. [CrossRef]

10. Gilmour, C.; Bell, T.; Soren, A.; Riedel, G.; Riedel, G.; Kopec, D.; Bodaly, D.; Ghosh, U. Activated carbon thin-layer placement as an in situ mercury remediation tool in a Penobscot River salt marsh. *Sci. Total Environ.* **2018**, *621*, 839–848. [CrossRef]
11. Ghosh, U.; Luthy, R.G.; Cornelissen, G.; Werner, D.; Menzie, C.A. In-situ sorbent amendments: A new direction in contaminated sediment management. *Environ. Sci. Technol.* **2011**, *45*, 1163–1168. [CrossRef]
12. Gomez-Eyles, J.L.; Yu, P.Q.; Beckingham, B.; Riedel, G.; Gilmour, C.C.; Ghosh, U. Evaluation of biochars and activated carbons for in situ remediation of sediments impacted with organics, mercury, and methylmercury. *Environ. Sci. Technol.* **2013**, *47*, 13721–13729. [CrossRef] [PubMed]
13. Ie, I.R.; Hung, C.H.; Jen, Y.S.; Yuan, C.S.; Chen, W.H. Adsorption of vapor-phase elemental mercury (Hg^0) and mercury chloride ($HgCl_2$) with innovative composite activated carbons impregnated with Na_2S and S^0 in different sequences. *Chem. Eng. J.* **2013**, *229*, 469–476. [CrossRef]
14. Puri, B.R.; Hazra, R.S. Carbon-sulphur surface complexes on charcoal. *Carbon* **1971**, *9*, 123–134. [CrossRef]
15. Li, Z.; Wu, L.; Liu, H.; Lan, H.; Qu, J. Improvement of aqueous mercury adsorption on activated coke by thiol-functionalization. *Chem. Eng. J.* **2013**, *228*, 925–934. [CrossRef]
16. Macıas-Garcıa, A.; Gomez-Serrano, V.; Alexandre-Franco, M.F.; Valenzuela-Calahorro, C. Adsorption of cadmium by sulphur dioxide treated activated carbon. *J. Hazard. Mater.* **2003**, *103*, 141–152. [CrossRef]
17. Ting, Y.; Chen, C.; Ch'ng, B.L.; Wang, Y.L.; Hsi, H.C. Using raw and sulfur-impregnated activated carbon as active cap for leaching inhibition of mercury and methylmercury from contaminated sediment. *J. Hazard. Mater.* **2018**, *354*, 116–124. [CrossRef]
18. Gong, Y.; Liu, Y.; Xiong, Z.; Kaback, D.; Zhao, D. Immobilization of mercury in field soil and sediment using carboxymethyl cellulose stabilized iron sulfide nanoparticles. *Nanotechnology* **2012**, *23*, 294007. [CrossRef]
19. Gong, Y.; Liu, Y.; Xiong, Z.; Zhao, D. Immobilization of mercury by carboxymethyl cellulose stabilized iron sulfide nanoparticles: Reaction mechanisms and effects of stabilizer and water chemistry. *Environ. Sci. Technol.* **2014**, *48*, 3986–3994. [CrossRef]
20. Han, D.S.; Orillano, M.; Khodary, A.; Duan, Y.; Batchelor, B.; Abdel-Wahab, A. Reactive iron sulfide (FeS)-supported ultrafiltration for removal of mercury (Hg (II)) from water. *Water Res.* **2014**, *53*, 310–321. [CrossRef]
21. Liu, J.; Valsaraj, K.T.; Devai, I.; DeLaune, R.E. Immobilization of aqueous Hg (II) by mackinawite (FeS). *J. Hazard. Mater.* **2008**, *157*, 432–440. [CrossRef]
22. Xiong, Z.; He, F.; Zhao, D.; Barnett, M.O. Immobilization of mercury in sediment using stabilized iron sulfide nanoparticles. *Water Res.* **2009**, *43*, 5171–5179. [CrossRef] [PubMed]
23. Paquette, K.; Helz, G. Solubility of cinnabar (red HgS) and implications for mercury speciation in sulfidic waters. *Water Air Soil Poll.* **1995**, *80*, 1053–1056. [CrossRef]
24. Benoit, J.M.; Gilmour, C.C.; Mason, R.P.; Heyes, A. Sulfide controls on mercury speciation and bioavailability to methylating bacteria in sediment pore waters. *Environ. Sci. Technol.* **1999**, *33*, 951–957. [CrossRef]
25. Ravichandran, M.; Aiken, G.R.; Ryan, J.N.; Reddy, M.M. Inhibition of precipitation and aggregation of metacinnabar (mercuric sulfide) by dissolved organic matter isolated from the Florida Everglades. *Environ. Sci. Technol.* **1999**, *33*, 1418–1423. [CrossRef]
26. Morse, J.W.; Arakaki, T. Adsorption and coprecipitation of divalent metals with mackinawite (FeS). *Geochim. Cosmochim. Acta* **1993**, *57*, 3635–3640. [CrossRef]
27. Wolthers, M.; Van der Gaast, S.J.; Rickard, D. The structure of disordered mackinawite. *Am. Mineral.* **2003**, *88*, 2007–2015. [CrossRef]
28. Hsi, H.C.; Rood, M.J.; Rostam-Abadi, M.; Chen, S.; Chang, R. Effects of sulfur impregnation temperature on the properties and mercury adsorption capacities of activated carbon fibers (ACFs). *Environ. Sci. Technol.* **2001**, *35*, 2785–2791. [CrossRef] [PubMed]
29. Hsu, C.J.; Chiou, H.J.; Chen, Y.H.; Lin, K.S.; Rood, M.J.; Hsi, H.C. Mercury adsorption and re-emission inhibition from actual WFGD wastewater using sulfur-containing activated carbon. *Environ. Res.* **2019**, *168*, 319–328. [CrossRef]
30. ASTM D6556-10. *Standard Test Method for Carbon Black—Total and External Surface Area by Nitrogen Adsorption*; ASTM International: West Conshohocken, PA, USA, 2012.
31. Lippens, B.C.; de Boer, J.H. Studies on pore systems in catalysts: V. the t method. *J. Catal.* **1965**, *4*, 319–323. [CrossRef]

32. Gee, G.W.; Bauder, J.W. Particle-size analysis. In *Methods of Soil Analysis, Part 1. Physical and Mineralogical Methods*, 2nd ed.; Klute, A., Ed.; Agronomy Monograph No. 9; American Society of Agronomy/Soil Science Society of America: Madison, WI, USA, 1986; pp. 383–411.
33. Nelson, D.W.; Sommers, L.E. Total carbon, organic carbon, and organic matter. In *Methods of Soil Analysis, Part 2. Chemical and Microbiological Properties*, 2nd ed.; Page, A.L., Ed.; Agronomy Series No. 9; American Society of Agronomy/Soil Science Society of America: Madison, WI, USA, 1982; pp. 539–579.
34. Lewis, P.A.; Klemm, D.J.; Lazorchak, J.M.; Norberg-King, T.J.; Peltier, W.H.; Heber, M.A. *Short-Term Methods for Estimating the Chronic Toxicity of Effluents and Receiving Waters to Freshwater Organisms*; US Environmental Protection Agency, Environmental Monitoring Systems Laboratory: Cincinnati, OH, USA, 1994.
35. Kester, D.R.; Duedall, I.W.; Connors, D.N.; Pytkowicz, R.M. Preparation of artificial seawater 1. *Limnol. Oceanogr.* **1967**, *12*, 176–179. [CrossRef]
36. Wang, Y.L.; Fang, M.D.; Chien, L.C.; Lin, C.C.; Hsi, H.C. Distribution of mercury and methylmercury in surface water and surface sediment of river, irrigation canal, reservoir and wetland in Taiwan. *Environ. Sci. Pollut. Res.* **2019**, *26*, 17762–17773. [CrossRef] [PubMed]
37. Kazemi, F.; Younesi, H.; Ghoreyshi, A.A.; Bahramifar, N.; Heidari, A. Thiol-incorporated activated carbon derived from fir wood sawdust as an efficient adsorbent for the removal of mercury ion: Batch and fixed-bed column studies. *Process Saf. Environ. Prot.* **2016**, *100*, 22–35. [CrossRef]
38. de Diego, A.; Tseng, C.M.; Dimov, N.; Amouroux, D.; Donard, O.F. Adsorption of aqueous inorganic mercury and methylmercury on suspended kaolin: Influence of sodium chloride, fulvic acid and particle content. *Appl. Organomet. Chem.* **2001**, *15*, 490–498. [CrossRef]
39. Ranganathan, K. Adsorption of Hg (II) ions from aqueous chloride solutions using powdered activated carbons. *Carbon* **2003**, *41*, 1087–1092. [CrossRef]
40. Thiem, L.; Badorek, D.; O'Connor, J.T. Removal of mercury from drinking water using activated carbon. *J. Am. Water Works Assoc.* **1976**, *68*, 447–451. [CrossRef]
41. Moreno, F.N.; Anderson, C.W.; Stewart, R.B.; Robinson, B.H.; Ghomshei, M.; Meech, J.A. Induced plant uptake and transport of mercury in the presence of sulphur-containing ligands and humic acid. *New Phytol.* **2005**, *166*, 445–454. [CrossRef]
42. Muller, K.A.; Brandt, C.C.; Mathews, T.J.; Brooks, S.C. Methylmercury sorption onto engineered materials. *J. Environ. Manag.* **2019**, *245*, 481–488. [CrossRef]
43. Schwartz, G.E.; Sanders, J.P.; McBurney, A.M.; Brown, S.S.; Ghosh, U.; Gilmour, C.C. Impact of dissolved organic matter on mercury and methylmercury sorption to activated carbon in soils: Implications for remediation. *Environ. Sci. Process. Impacts* **2019**, *21*, 485–496. [CrossRef]
44. Johs, A.; Eller, V.A.; Mehlhorn, T.L.; Brooks, S.C.; Harper, D.P.; Mayes, M.A.; Pierce, E.M.; Peterson, M.J. Dissolved organic matter reduces the effectiveness of sorbents for mercury removal. *Sci. Total Environ.* **2019**, *690*, 410–416. [CrossRef]
45. Skyllberg, U.; Bloom, P.R.; Qian, J.; Lin, C.M.; Bleam, W.F. Complexation of mercury (II) in soil organic matter: EXAFS evidence for linear two-coordination with reduced sulfur groups. *Environ. Sci. Technol.* **2006**, *40*, 4174–4180. [CrossRef]
46. Skyllberg, U.; Drott, A. Competition between disordered iron sulfide and natural organic matter associated thiols for mercury (II): An EXAFS study. *Environ. Sci. Technol.* **2010**, *44*, 1254–1259. [CrossRef] [PubMed]
47. Pandey, A.K.; Pandey, S.D.; Misra, V. Stability constants of metal–humic acid complexes and its role in environmental detoxification. *Ecotoxicol. Environ. Saf.* **2000**, *47*, 195–200. [CrossRef]
48. Ting, Y.; Ch'ng, B.L.; Chen, C.; Ou, M.Y.; Cheng, Y.H.; Hsu, C.J.; Hsi, H.C. A simulation study of mercury immobilization in estuary sediment microcosm by activated carbon/clay-based thin-layer capping under artificial flow and turbation. *Sci. Total Environ.* **2020**, *708*, 135068. [CrossRef] [PubMed]
49. Compeau, G.C.; Bartha, R. Effect of salinity on mercury-methylating activity of sulfate-reducing bacteria in estuarine sediments. *Appl. Environ. Microbiol.* **1987**, *53*, 261–265. [CrossRef] [PubMed]
50. Ullrich, S.M.; Tanton, T.W.; Abdrashitova, S.A. Mercury in the aquatic environment: A review of factors affecting methylation. *Crit. Rev. Environ. Sci. Technol.* **2001**, *31*, 241–293. [CrossRef]

Article

Using Mixed Active Capping to Remediate Multiple Potential Toxic Metal Contaminated Sediment for Reducing Environmental Risk

Meng-Yuan Ou [1], Yu Ting [1], Boon-Lek Ch'ng [1], Chi Chen [1], Yung-Hua Cheng [1], Tien-Chin Chang [2] and Hsing-Cheng Hsi [1,*]

1 Graduate Institute of Environmental Engineering, National Taiwan University, No. 1, Sec. 4, Roosevelt Rd., Taipei 10671, Taiwan; auanlie@outlook.com (M.-Y.O.); yuting821216@gmail.com (Y.T.); r06541135@ntu.edu.tw (B.-L.C.); q2461015@gmail.com (C.C.); r08541121@ntu.edu.tw (Y.-H.C.)

2 Institute of Environmental Engineering and Management, National Taipei University of Technology, No. 1, Sec. 3, Zhongxiao E. Rd., Taipei 10608, Taiwan; tcchang@ntut.edu.tw

* Correspondence: hchsi@ntu.edu.tw; Tel.: +886-2-33664374

Received: 18 May 2020; Accepted: 27 June 2020; Published: 1 July 2020

Abstract: In this study, kaolinite, carbon black (CB), iron sulfide (FeS), hydroxyapatite (HAP), and oyster shell powder (OSP) were selected as potentially ideal amendments to immobilize metals in sediment, including Ni, Cr, Cu, Zn, and Hg. In aqueous batch experiments, the five adsorbents were tested for capturing the five potential toxic metals individually at various concentrations. HAP and OSP showed the largest removal efficiencies towards Ni (OSP: 76.47%), Cr (OSP: 100.00%), Cu (HAP: 98.39%), and Zn (HAP: 64.56%), with CB taking the third place. In contrast, FeS and CB played a more significant role in Hg removal (FeS: 100.00%; CB: 86.40%). In the modified six-column microcosm experiments, five mixing ratios based on various considerations using the five adsorbent materials were tested; the water samples were collected and analyzed every week for 135 days. Results showed that caps including CB could immobilize the release of Hg and methylmercury (MeHg) better than those with FeS. More economical caps, namely, with a higher portion of OSP in the mixed capping, could not reach comparable effects to those with more HAP for immobilizing Ni, but performed almost the same for the other four metals. All columns with active caps showed greater metal immobilization as compared to the controlled column without caps.

Keywords: active capping; toxic metal; sediment; remediation; multiple materials

---

1. Introduction

Wastewater containing potential toxic metals originating from anthropogenic activities discharged to river streams is a widespread environmental issue nowadays, processing significant toxicity to aquatic organisms and accumulating by food chain, finally causing various diseases and disorders [1]. Natural processes are frequently inadequate to deal with the elevated metal loading, therefore there is an urgent need for remediation measures [2]. Thin-layer capping has been applied as an economically-feasible in-situ method for sediment remediation, reducing contaminants release from sediment to overlying water, subsequently reducing ecological and human health risk [3]. Sediment can be seen as an important sink of various organic and inorganic compounds, resulting in the simultaneous existence of several different contaminants [4–6]. However, most of the previous studies focused on only one or two capping materials. To cope with highly complex conditions in sediment, mixed capping with multiple materials was proposed. Notably, using cheap and effective alternatives for the removal of potential toxic metals could reduce operating costs, reduce the prices of products, improve competitiveness, and benefit the environment [7].

Several relatively cheap, environment-friendly materials have been proposed and examined. Based on previous research [7–22], kaolinite, carbon black (CB), iron sulfide (FeS), oyster shell powder (OSP), and hydroxyapatite (HAP; $Ca_{10}(PO_4)_6(OH)_2$,) were considered as potentially ideal amendments to be part of the mixed caps. Previous reports provided excellent results showing that kaolinite can be used as a cheap and naturally occurring adsorbent to remove Pb^{2+}, Cd^{2+}, Ni^{2+}, Co^{2+}, Cr^{6+}, Zn^{2+}, and Cu^{2+} from aqueous solution in both single and multi-metal ions [8–11]. Sulfide minerals are shown to be ideal materials to scavenge Hg^{2+} by immobilizing it through adsorption or co-precipitation as a discrete sulfide phase (HgS) and solid solution formation with iron sulfides [12,13]. HAP jumped out recently because of its high adsorption capacity of various metals, which brings the perspective for removal of Cd^{2+}, Zn^{2+}, Pb^{2+}, U^{6+}, Co^{2+} ions, and so on [14–17]. HAP is a nanostructured material, which was recently synthesized from some high-calcium biological wastes as raw materials, such as seashells and eggshells [18]. In the hunt of such low-cost and efficient raw materials for the production of HAP, OSP has emerged as a suitable one. As for OSP itself, owing to its low price and basicity, it has been employed widely for stabilization/solidification of many kinds of potential toxic metals (As, Pb, Ni, and Cu, etc.) [7,19–22]. The solution pH of OSP greatly affected the adsorption process towards Cu^{2+}, with an optimum adsorption pH of 5.5 and an overall negative surface charge facilitating the adsorption process [19]. The adsorption capacities of OSP towards Cu^{2+} and Ni^{2+} could reach 49.26–103.1 mg g^{-1} and 48.75–94.3 mg g^{-1}, respectively, through physical adsorption mechanisms [7].

The present study aims to find mixed amendments with the best comprehensive benefits for immobilizing potential toxic metals in contaminated sediment. Based on the previous works, kaolinite, CB, FeS, OSP, and HAP were selected as potentially ideal amendments to immobilize metals in sediment. According to the survey report of Taiwan in 2012 [23] on main rivers with contaminated sediment, Ni, Cr, Cu, Zn, and Hg are the five major potential toxic metals present in sediments, which were selected as the targets of remediation in this study. Furthermore, methylmercury (MeHg) converted from Hg was measured in this study due to its high toxicity and bioaccumulative ability. Aqueous batch experiments were first conducted by using the five adsorbents to capture the five potential toxic metals individually at various concentrations. Then, five columns containing contaminated sediment covered by different mixed materials were set based on the obtained results of batch experiments to better comprehend the immobilization effectiveness of mixed capping for the potential toxic metals as compared to the controlled column without caps. Results from this study are helpful in designing an economically and technically feasible in−situ approach for sediment remediation, which could not only reduce the potential toxic metal release from sediment to overlying water, but also reduce the ecological and human health risk.

2. Materials and Methods

2.1. Adsorbents Preparation

The kaolinite and FeS used in this study were all reagent-grade chemicals purchased from Sigma-Aldrich. The CB used was obtained from Enrestec Inc. (Tainan, Taiwan), and was considered as a low-cost, recycled materials because it was a byproduct from waste rubber tire pyrolysis for oil production.

OSP was prepared by grinding oyster shells collected as food waste materials. The oyster shells were scrubbed carefully to remove impurities, then dried for 48 h at 100 °C [19] and ground to homogenized powder that could pass through a 30-mesh sieve. Then this sieved material was shaken over a 60-mesh sieve and any material that passed through was rejected (ASTM D2765). [24]

To synthesize HAP, the OSP passed though the 60-mesh sieve was collected as a raw material. The synthesis was operated according to the following reaction: $10Ca^{2+} + 6PO_4^{3-} + 2OH^- = Ca_{10}(PO_4)_6(OH)_2$, referring to previous studies [18,25–27] (Figure 1). In the synthesis process, an amount of Na_2HPO_4 and the corresponding amount of oyster shell powder (Ca/P molar ratio = 1.67) were used as P and Ca precursors, respectively. Firstly, 5 g of that oyster shell powder (<60 mesh) was dissolved in 1:3 hydrochloride acid/water solution and stirred thoroughly. The supernatant of the solution filtered by 0.45 μm filter was put into 500 mL of 0.1 M EDTA. Then reaction process was carried out by drop wise addition of 0.06 M Na_2HPO_4 solution under continuous stirring of 200 rpm. The pH value of the reaction mixture was regulated within the range of 10.5–11.5 by 5 M NaOH solution and maintained throughout the process the dripping velocity was controlled at 2 mL min^{-1}. After mixing, the resulting mixture was left to maturate for the next 24 h. The obtained white precipitate was washed with distilled water by centrifuge and dried in the hot air oven at 65 °C for 24 h to get the final product.

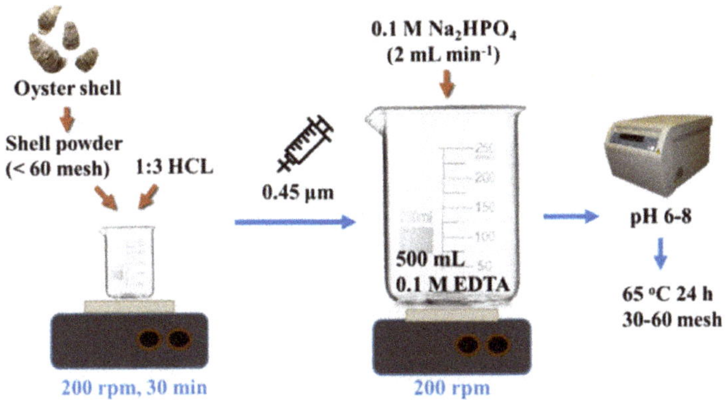

Figure 1. The process of synthesizing HAP.

All of the above materials were sieved to the range of 30–60 mesh (ASTM D2765) [24] and analyzed using a physisorption analyzer (Micromeritics Inc. ASAP 2420, Norcross, GA, USA) to obtain the 77 K N_2 adsorption isotherm. The total surface area was then obtained based on the Brunauer-Emmett-Teller (BET) equation and described as BET surface area (S_{BET}). The chemical compositions of the five materials were analyzed for the C/H/N/S contents (Elementar Vario EL cube, Langenselbold, Germany) and O content by automatic elemental analyzers (Flash 2000, Thermo Fisher Scientific, Waltham, MA, USA).

2.2. *Artificial Fresh Water and Sediment Incubation*

In order to simulate the river environment, artificial freshwater prepared based on the formula provided by USEPA [28] was used in this system, instead of using purified or deionized water. To prepare 20 L of synthetic, moderately hard, reconstituted fresh water, the reagent grade chemicals as follows were used: 1.20 g $MgSO_4$, 1.92 g $NaHCO_3$, 0.080 g KCl, and 1.06 g $CaSO_4$.

The metal incubation concentrations of artificial sediment and reagent grade chemicals were determined with reference to the survey report of Taiwan in 2012 [23] on main rivers with contaminated sediment. The sediment was designed to consist of 400 mg kg^{-1} Ni, 400 mg kg^{-1} Cr, 1000 mg kg^{-1} Cu, 1000 mg kg^{-1} Zn, and 50 mg kg^{-1} Hg by adding $Ni(NO_3)_2{\cdot}6H_2O$, $Cr(NO_3)_3{\cdot}9H_2O$, $Cu(NO_3)_2{\cdot}3H_2O$, $Zn(NO_3)_2{\cdot}6H_2O$, and $HgCl_2$. The sediment (2000 g) was placed in a 3 L glass bottle, filled with artificial freshwater, mixed thoroughly, sealed tightly, and put aside for incubation. The time of incubation lasted up to 60 days.

2.3. *Aqueous Batch Experiments*

The aqueous adsorption experiment was performed according to Bouhamed et al. [1]. To determine the adsorption isotherms of adsorbents, 10, 30, and 50 mg L^{-1} of Ni, Cr, Cu, and Zn were tested by aqueous batch experiments. For Hg, 0.2, 0.6, and 1.0 mg L^{-1} were tested. A serial dilution of standard solution was made to the intended concentrations; the pH value of the solution was controlled to be 5 ± 0.1 by using 0.01 M NaOH and 0.01 M HCl, which simulates the real contaminated sediment environment.

Batch adsorption experiments were carried out in a rotary shaker at 125 rpm using 50 mL capped glass bottles containing 20 mL of metal ion solutions and 25 mg of adsorbents, similar to our previous study [29]. The temperature was controlled at 30 °C and the contact time was up to 24 h to achieve an adsorption equilibrium. All the experimental operations were triplicated.

After shaking, samples were filtered with 0.45 μm filters and the supernatant was kept for metal analyses. For a long-time preservation, each sample of Ni, Cr, Cu, and Zn was preserved with 0.5% HNO_3 and measured by flame atomic absorption spectroscopy (FAAS; Perkin Elmer AAnalyst 800, Waltham, MA, USA); each sample of Hg was preserved with 0.5% BrCl and estimated by cold vapor atomic fluorescence spectroscopy (CVAFS; Brooks Rand Automated Total Mercury System, Seattle, WA, USA). QA/QC were regularly checked by analyzing duplicate samples and quality control samples from each batch.

2.4. Microcosm Experiments

2.4.1. Microcosm Design

The microcosm shown in Figure 2 was designed by modifying the system used in Ting et al. [29]. Multi-columns containing contaminated sediment with mixed caps on the top were set to stimulate the release of metal compounds and examine the efficiencies of capping materials. Water entered the system from the bottom of the column, vertically moved upwards to fill the column, and then discharged to the outside of the system.

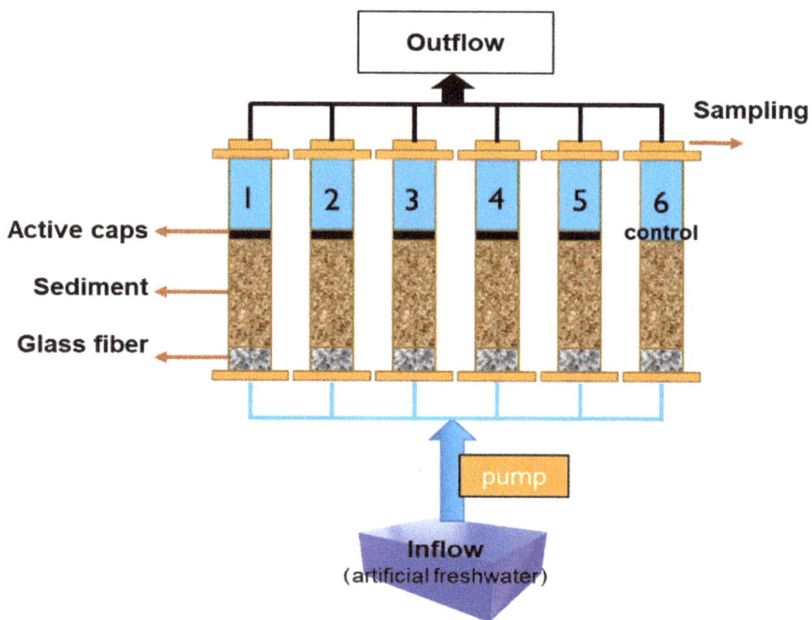

Figure 2. The lab-scale microcosm system.

2.4.2. Microcosm Operation

Six freshwater columns were set up to investigate the efficiencies and stabilities of different mixed caps. Dried incubated sediment was added into the columns and each column contained 200 g of sediment. Five columns were capped by the mixed adsorbents with five different ratios, which were determined according to the results of batch experiments. The sixth column was not capped as the controlled unit. To start up the system, each column was filled with water and allowed to settle for 24 h. Next, the microcosm was activated to start inflow and this day was counted as the operation day 1. The flow rate was maintained to be 0.2 L day^{-1} constantly by using a peristaltic pump (Gamma ST100SV2, Shakopee, MN, USA).

Notably, in order to investigate the stabilization of different mixed caps, the pH of artificial water was adjusted to 3 ± 0.1 by adding HNO_3 during day 101–114 to simulate the extreme condition with acid influent.

2.4.3. Water Sampling and Analyses

Periodic water sampling was conducted continuously once a week for up to two months. At each sampling time, the pH, dissolved oxygen (DO), electrical conductivity (EC), oxidation-reduction potential (ORP) were measured one by one initially. DO of the overlying water was directly measured by a DO meter (EXStik DO600, Nashua, NH, USA). pH value was measured by a pH meter (SunTex SR-2300, Hsi-Chih, Taipei, Taiwan). The electrical conductivity (EC) was measured by a conductivity meter (Taina EZDO-6021, Taiwan). The oxidation–reduction potential (ORP) of sediment within the depth of 1–3 cm of caps was measured by an ORP meter (SunTex SR-2300, Hsi-Chih, Taipei, Taiwan). After the basic measurement, 100 mL of water was collected from each column by disposable syringes and filtered with 0.45 μm filter membranes. The supernatant was kept in 20 mL glass bottles in a refrigerator at 4 °C before analyses.

The samples prepared for Hg analysis were preserved with 0.5% BrCl. For Ni, Cr, Cu, and Zn analyses, samples were preserved with 0.5% HNO_3 and determined by inductively coupled plasma optical emission spectrometry (Agilent ICP-OES 700 series, Santa Clara, CA, USA). For MeHg analysis, samples were put in brown bottles in the shadows to avoid light and analyzed immediately by ethylation, purge and trap, and gas chromatography/CVAFS (Brooks Rand Automated Total Mercury System, Seattle, WA, USA). Total organic carbon (TOC) was analyzed by a TOC analyzer (OI Analytical Aurora 1030W, College Station, TX, USA). QA/QC were regularly checked by analyzing the quality control samples from each batch.

2.4.4. Statistical Analysis

A one-way ANOVA, followed by a least significant difference (LSD) test ($p<0.05$), was used to determine the significance differences among columns (IBM SPSS statistics).

3. Results

3.1. *Adsorbents Properties*

Five capping adsorbents, sieved to 30–60 mesh, were analyzed and their physical and chemical properties are summarized in Table 1, including the BET surface areas (S_{BET}), average pore sizes, and total pore volumes (V_{total}). The data revealed that HAP had the absolutely largest BET surface area and pore volume among these five materials, and CB had a relatively larger specific surface area and the second largest pore volume.

Table 1. BET surface areas, average pore sizes and total pore volumes (V_{total}) of five materials.

Adsorbent	S_{BET} (m^2 g^{-1})	Pore Size (nm)	V_{total} (cm^3 g^{-1})
Kaolinite	23.3	14.0	0.082
FeS	106.2	60.2	0.042
CB	93.0	17.3	0.376
HAP	367.4	9.2	0.501
OSP	0.06	4.8	0.011

The results of element analyses of the five materials are given in Table 2. FeS was not analyzed and theoretically it is composed of 63.64 wt% of iron and 36.36 wt% of sulfur. It can be seen that CB was rich in C (77.87 wt%) and had a great amount of S (2.53 wt%) because it originated from waste tire. OSP and HAP were rich in O (32.29 and 13.24 wt%, respectively). The proportions of C in OSP roughly corresponded to the proportion of $CaCO_3$ in oyster shell.

Table 2. Elemental analyses of the five test materials.

Adsorbent	C (wt%)	N (wt%)	H (wt%)	O (wt%)	S (wt%)
Kaolinite	0.03	0.15	1.43	10.88	0.05
FeS			-		
CB	77.87	0.46	1.02	1.6	2.53
HAP	0.00	0.11	1.71	13.24	0.00
OSP	11.61	0.20	0.55	32.29	0.07

3.2. Adsorption Efficiency

The aqueous metal removal efficiency by the adsorbents was calculated with the following equation:

$$R = \frac{C_0 - C_t}{C_0} \times 100\%, \quad (1)$$

where R (%) is the removal efficiency of adsorbents, C_0 (mg L^{-1}) is the initial metal concentration detected in blank solution, and C_t (mg L^{-1}) is the concentration of remaining sorbate at any time.

Figure 3 and Table 3 show the removal efficiencies of five materials towards five metals at various initial concentrations, which illustrated roughly that OSP and HAP had the best affinities to Cr, Zn, Cu, and Ni, and CB took the third place. As for Cu, HAP performed the largest removal efficiency than the others with a reduction of 98.39%. For Ni, OSP presented a significant removal rate of 76.47% at a low initial concentration and, contrarily, poor efficiency at high initial metal concentrations. For the adsorption of Hg, inversely, FeS showed the best removal efficiency, which was up to 100%. CB was the second excellent one with a removal rate of 86.4%.

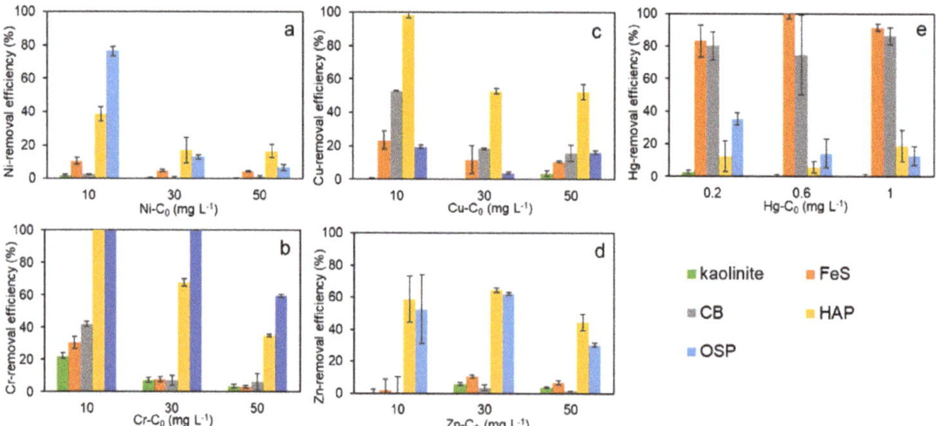

Figure 3. Removal efficiencies of five materials towards (**a**) Ni, (**b**) Cr, (**c**) Cu, (**d**) Zn, and (**e**) Hg at various initial concentrations.

Table 3. Removal efficiencies (in percentage) of five adsorbents for five metals (n = 3).

Metal	C_0 (mg L^{-1})	Kaolinite	FeS	CB	HAP	OSP
Ni	10	2.04 ± 0.50	10.73 ± 1.98	2.60 ± 0.27	38.85 ± 4.36	76.47 ± 2.73
	30	0.86 ± 0.12	4.96 ± 0.91	0.60 ± 0.53	17.11 ± 7.56	12.99 ± 1.19
	50	0.41 ± 0.15	4.63 ± 0.48	1.37 ± 0.72	16.62 ± 3.98	6.85 ± 1.81
Cr	10	22.15 ± 1.66	30.65 ± 3.85	42.00 ± 1.41	102.58 ± 0.15	103.13 ± 0.03
	30	7.38 ± 1.61	7.88 ± 1.63	6.98 ± 3.11	67.81 ± 2.30	105.18 ± 0.07
	50	3.73 ± 1.10	3.07 ± 0.77	6.19 ± 5.09	35.06 ± 0.73	59.58 ± 0.90
Cu	10	−1.13 ± 0.95	23.45 ± 5.41	52.80 ± 0.21	98.39 ± 1.65	19.43 ± 1.17
	30	−0.20 ± 0.11	11.89 ± 8.29	18.27 ± 0.41	52.90 ± 1.85	3.82 ± 0.76
	50	3.69 ± 1.83	10.78 ± 0.69	15.91 ± 4.83	52.41 ± 4.46	16.05 ± 1.14
Zn	10	−6.63 ± 2.65	2.09 ± 6.73	−29.32 ± 10.30	58.88 ± 14.26	52.71 ± 21.53
	30	5.86 ± 0.81	10.55 ± 1.03	3.47 ± 1.62	64.56 ± 1.20	62.30 ± 0.98
	50	3.69 ± 0.45	6.71 ± 1.33	1.44 ± 0.00	44.38 ± 5.02	30.62 ± 1.21
Hg	0.2	2.93 ± 1.23	83.37 ± 9.65	80.30 ± 8.55	12.48 ± 9.35	35.32 ± 3.81
	0.6	−2.36 ± 1.15	101.12 ± 3.15	74.70 ± 24.47	5.92 ± 3.47	14.43 ± 8.85
	1	−0.50 ± 1.75	91.51 ± 2.10	86.40 ± 5.09	18.86 ± 9.43	12.86 ± 5.70

3.3. Microcosm

3.3.1. Mixed Caps Design

Five columns covered by different mixed materials and one controlled column were set as shown in Table 4 for conducting the microcosm experiments. The design of capping mixing ratio was based on the integrating consideration of removal efficiencies, cost of preparation, and utilization of recycling resources.

Table 4. Six columns covered by different mixed materials.

Column	1	2	3	4	5	6
kaolinite (wt%)	10	10	10	10	-	
FeS (wt%)	2.5	-	5	2.5	2.5	
CB (wt%)	2.5	5	-	2.5	2.5	-
HAP (wt%)	35	35	35	5	35	
OSP (wt%)	50	50	50	80	50	
Total (wt%)	100	100	100	100	90	

Based on the aforementioned results from batch experiments, HAP and OSP were selected as the major materials to adsorb Ni, Cr, Cu, and Zn; FeS and CB were selected as the major materials to immobilize total Hg (THg) and MeHg. Although the adsorption effect of kaolinite on these five metals was not significant, it has been shifted that kaolinite may probably be more capable of stabilizing the caps than other clays [30] and was therefore added for the purpose of resisting flow disturbance. Column 1 was designed as the ideally best column with appropriate amounts of each material. Column 2 replaced FeS completely with CB and Column 3 did the opposite for comparing FeS with CB. As far as costs and complexity of preparation were concerned, Column 4 tried to reduce the proportion of HAP to grope for a cheaper mixed amendment with equal ability. The cap ratio of Column 5 was almost the same as Column 1, except for the absence of 10 wt% kaolinite, which was designed to investigate the necessity of existence of kaolinite. The last column was composed of only sediment without caps as the controlled group.

3.3.2. Results of Sediment Incubation

The concentrations of the five metals before and after 60 days of incubation and the concentrations of the five metals in the supernatant after incubation are shown in Table 5. Concentrations of metals in the sediment after incubation had almost met the experimental requirements. It is worth mentioning that the concentrations of Ni and Zn of supernatant were much higher than those of the other three metals.

Table 5. Concentrations of the five metals in sediment before and after 60 days of incubation and the concentrations of the five metals in the supernatant after incubation.

Metal	Sediment (mg kg^{-1})			Supernatant (mg L^{-1})
	Before	Design	After	
Ni	61.35 ± 1.46	400	519.59 ± 9.52	133.46 ± 0.98
Cr	102.16 ± 0.63	400	593.96 ± 2.81	0.16 ± 0.002
Cu	94.64 ± 2.38	1000	1383.30 ± 6.58	17.71 ± 0.12
Zn	373.70 ± 12.18	1000	1579.13 ± 14.31	358.79 ± 1.06
THg	0.20 ± 0.03	50	72.32 ± 1.96	0. 03 ± 0.008
MeHg	$1.50 \times 10^{-4} \pm 2.54 \times 10^{-5}$	–	$8.56 \times 10^{-3} \pm 3.23 \times 10^{-3}$	–

3.3.3. pH and ORP

Figure 4 shows the changes of oxidation-reduction potential (ORP) and water pH in the microcosm during the operation. The data of changes of electrical conductivity (EC), pH, dissolved oxygen (DO), and oxidation-reduction potential (ORP) are also shown in Tables S1–S4. Higher pH values and lower ORP in columns capped with materials were attained as compared to the one without caps. However, the differences tended to be slight during the later stage.

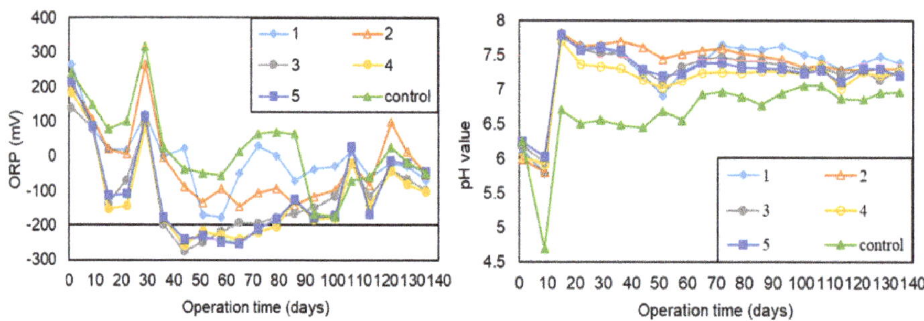

Figure 4. The changes of oxidation-reduction potential (ORP) and water pH values in microcosm during the operation.

3.3.4. Metal Immobilization

The concentrations of five metals and MeHg are shown in Figure 5 and Tables S5–S10 (in Supplementary Material). The results of one-way ANOVA and LSD test ($p < 0.05$) are shown in Tables 6 and 7. In consistence with the expectation, Column 1 showed the most considerable comprehensive consequences of inhibiting the release of metals to the overlying water. It was unexpected that even without 10 wt% of kaolinite, Column 5 achieved almost the same results as Column 1, which is also proved by no significant ($p > 0.05$) correlation in ANOVA (Table 6) between the two columns for all metals, indicating that kaolinite did not show greater stabilization ability for mixed caps as compared to the other test materials.

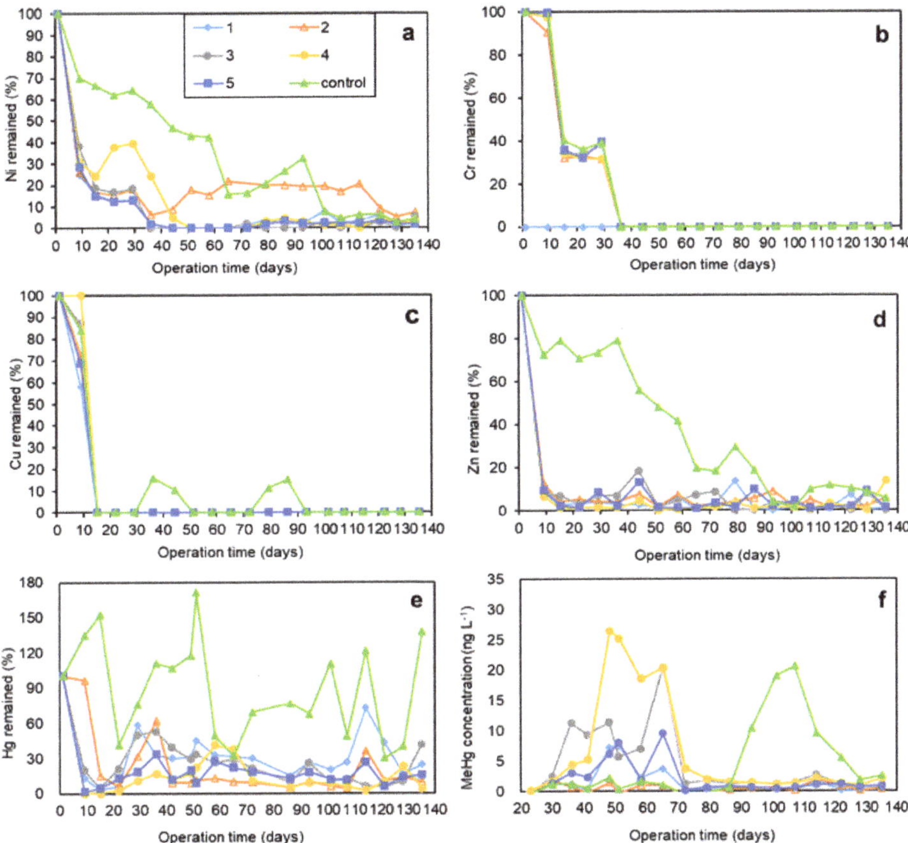

Figure 5. The remaining proportion (%) of (**a**) Ni, (**b**) Cr, (**c**) Cu, (**d**) Zn, (**e**) Hg, and (**f**) the concentration of MeHg in six columns. The remaining proportion in aqueous phase was calculated through dividing the concentration of each contaminant by the concentration on the first day.

Table 6. Metal concentrations in six columns before adding acid (day 1–100).

Column	Ni (n = 14) (mg L^{-1})	Cr (n = 14) (mg L^{-1})	Cu (n = 14) (mg L^{-1})	Zn (n = 14) (mg L^{-1})	THg (n = 14) (μg L^{-1})	MeHg (n = 12) (ng L^{-1})
1	0.091 ± 0.189^{b}	0.000 ± 0.000^{a}	0.013 ± 0.036^{a}	0.150 ± 0.354^{b}	0.145 ± 0.109^{c}	2.09 ± 2.68^{bc}
2	0.162 ± 0.157^{b}	0.015 ± 0.025^{a}	0.012 ± 0.030^{a}	0.140 ± 0.289^{b}	0.382 ± 0.463^{ab}	0.535 ± 0.622^{c}
3	0.064 ± 0.126^{b}	0.015 ± 0.025^{a}	0.010 ± 0.027^{a}	0.090 ± 0.182^{b}	0.233 ± 0.166^{bc}	6.06 ± 6.09^{ab}
4	0.140 ± 0.200^{b}	0.016 ± 0.026^{a}	0.012 ± 0.031^{a}	0.131 ± 0.385^{b}	0.288 ± 0.373^{abc}	9.18 ± 10.2^{a}
5	0.079 ± 0.165^{b}	0.015 ± 0.025^{a}	0.012 ± 0.031^{a}	0.135 ± 0.311^{b}	0.178 ± 0.194^{bc}	2.85 ± 3.25^{bc}
6	0.315 ± 0.161^{a}	0.015 ± 0.025^{a}	0.018 ± 0.035^{a}	0.498 ± 0.288^{a}	0.496 ± 0.222^{a}	1.56 ± 2.85^{c}

Different letters for metal concentrations in six columns indicate a significant difference at $p < 0.05$. For example, the concentrations of Ni in Column 1 with letter b are significantly different from the concentrations of Ni in Column 6 with letter a.

When comparing Column 1 (50% OSP + 35% HAP) with Column 4 (80% OSP + 5% HAP), it is demonstrated by Figure 5 that the releases of Ni, Cr, Cu, and Zn were inhibited well in both columns. However, Column 4 with less proportion of HAP could not achieve an equal effect to Column 1 for Ni, Cr and MeHg immobilization (Figure 5), and a significant correlation was observed by ANOVA ($p < 0.05$) (Table 6) in MeHg concentrations of the two columns (Table 6).

Table 7. Metal concentrations in six columns after adding acid (day 101–135)

Column	Ni (n = 6) (mg L^{-1})	Zn (n = 6) (mg L^{-1})	THg (n = 6) (μg L^{-1})	MeHg (n = 6) (ng L^{-1})
1	0.029 ± 0.017^{ab}	0.024 ± 0.038^{b}	0.151 ± 0.094^{b}	0.453 ± 0.544^{b}
2	0.029 ± 0.017^{ab}	0.033 ± 0.022^{ab}	0.182 ± 0.162^{b}	0.605 ± 0.672^{b}
3	0.009 ± 0.010^{c}	0.017 ± 0.023^{b}	0.100 ± 0.096^{b}	1.29 ± 0.729^{b}
4	0.016 ± 0.012^{bc}	0.049 ± 0.076^{ab}	0.121 ± 0.106^{b}	1.42 ± 0.532^{b}
5	0.015 ± 0.005^{bc}	0.038 ± 0.039^{ab}	0.113 ± 0.057^{b}	0.721 ± 0.311^{b}
6	0.035 ± 0.011^{a}	0.078 ± 0.036^{a}	0.433 ± 0.251^{a}	9.80 ± 8.20^{a}

Different letters for metal concentrations in six columns indicate a significant difference at $p < 0.05$. For example, the concentrations of Ni in Column 1 with letter a and b are significantly different from the concentrations of Ni in Column 3 with letter c.

To compare FeS with CB, 5% FeS in Column 3 made it more effective in immobilizing Ni than 5% CB in Column 2. On the other hand, Column 3 did worse than Column 2 for lowering the concentration of MeHg in the long term, which was demonstrated doubly by Figure 5 and statistical analysis (Table 6).

The stabilization of different mixed caps under extreme condition with acid influent during day 101–114 is also shown in Figure 5. It was observed that the concentrations of Hg and MeHg in every column increased during the days with acid influent (Figure 5f), especially for the controlled column without capping (Table 7). Also, Column 6 released the most amount of Zn during day 101–114 (Figure 5d). The experimental results also suggested that columns with caps were less affected by acid influent than the controlled one, which could be supported by strong significant correlations observed by ANOVA between Column 6 and the other columns.

3.3.5. Total Organic Matter (TOC)

Figure 6 shows that the amount of TOC was positively related to the concentration of MeHg. Column 4, with the largest percentage of OSP, released maximum MeHg during day 20–80. In a later stage from approximately day 85, however, both MeHg and TOC concentrations in Column 6 increased suddenly.

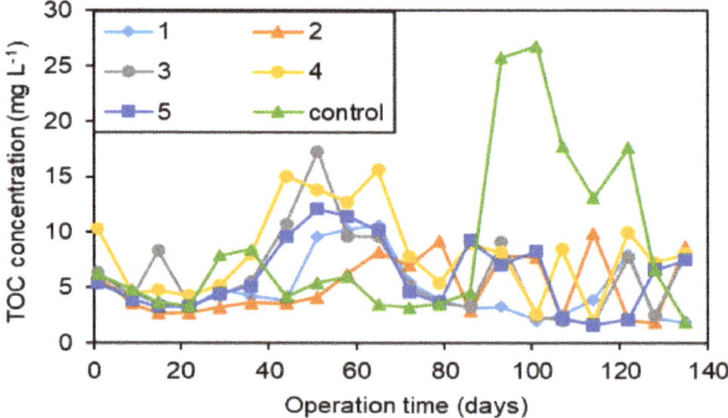

Figure 6. Total organic carbon (TOC) in six columns during the operation

4. Discussion

The data of material properties showed the absolutely largest BET surface area and pore volume of HAP, CB, and FeS, which may lead to its excellent performance during adsorption due to more adsorption sites. In contrast, OSP is a non-porous material with insignificant surface area and pore volume, but OSP still had appreciable adsorption performance for several metals, suggesting that the adsorption behavior of OSP is different with its derived HAP. Abundant C content in CB indicated its additional benefits as an active material [31]. HAP was rich in H and O, probably as a result of the richness of hydroxyl groups on its surface [32], contributing to its high ability as a metal scavenger [33].

In the part of batch experiments, OSP and HAP had the optimum removal efficiencies for Cr, Zn, Cu, and Ni. HAP is regarded as an ideal material with large specific surface and high stability under both reducing and oxidizing conditions [16]. It can release phosphate to interact with metals, forming metal phosphates of low water solubility [34]. OSP has been proven to be able to raise the pH value when applied to soil [35]. If the pH is above the value that promotes metal precipitation, the removal mechanisms will be related to precipitation [36]. However, it is known that the hydroxides of Ni have the highest solubility product (K_{sp}) among those of the five metals, making it comparatively difficult for Ni to be removed by precipitation. As for Hg, FeS and CB showed the best immobilizing ability due to their high Hg affinity. Notably, the CB used in this study was a waste tire recycled product, which contains a significant amount of S that can form chemical bonding with Hg.

The metal potential leachability could determine their environmental risks, and it has been reported to decrease in the order of Zn > Ni > Cu > Cr at pH 4 [37], which corresponded to the concentrations of the five metals in the supernatant after incubation, revealed stronger leachability of Ni and Zn, and was related to their behaviors in the microcosm.

It is generally accepted that sediment pH is an important factor in the adsorption of adsorbate on adsorbent. Under low pH condition, most metal ions are in the cationic state in the solution. The hydrogen ions can compete for the adsorption sites with the metal ions, influencing the exchange adsorption of potential toxic metals, promoting the desorption of metal ions and causing a higher release rate [38,39]. The ability of OSP to raise the concentration of hydroxide ions mentioned before may account for the increase of pH in the columns and its high removal capacity of metals. As one of the most important factors influencing the mobility of metals, the increase of ORP in sediment will correspondingly promote the oxidization of metal sulfides and the degradation of organic compounds, both accelerating the release of the adsorbed/complexing metals [40]. It appeared that ORP of Column 6 was higher than the other five capped in general, mainly due to the decrease of dissolved oxygen caused by caps in the other five columns.

In the part of the lab-scale microcosm experiments, the data from Column 5 indicated the unnecessary existence of kaolinite whether for adsorption or for stabilization in such a condition, which is inconsistent with the suggestions from some earlier studies. The reason may be due to that HAP and OSP could play the same role as kaolinite to stabilize the caps. The difference between Column 1 and Column 4 shown in Figure 5 was related to different performances of HAP and OSP to remove Ni as Figure 3a illustrated. Nevertheless, using caps mainly composed of OSP can yet be regarded as a good choice when considering costs. In batch adsorption, FeS showed the third best adsorption ability for Ni by a reduction of 10.73%, and yet CB hardly worked (Figure 3a). Thus, the release in the later stage of the experiment of Ni in Column 2 may be due to the ineffectiveness of CB. In contrast, Column 3 (5% FeS) did worse than Column 2 (5% CB) for lowering the concentration of MeHg in the long term (Figure 3f). FeS can reduce the concentration of soluble Hg species, resulting in less methylation of Hg [41]. Moreover, part of FeS was likely to be converted into insoluble HgS and taken out by the overlying water, which may explain the unsuccessful results of Column 3 to immobilize MeHg.

The relationship between TOC and MeHg during day 20–80 was likely caused by the remaining organic matter existing in OSP. It has been observed that organic matter content seemed to play a critical role for MeHg formation, acting as electron donor for Hg methylation bacteria [42,43]. The sediment in Column 4 with more OSP would likely release a greater amount of MeHg. Without capping, TOC was possibly easier to release from sediment, which then caused the uprush of MeHg in Column 6 from day 85.

5. Conclusions

Based on the results obtained from the aqueous batch experiments, OSP and HAP had the optimum removal efficiencies to Cr, Zn, Cu, and Ni with CB taking the third place. As for Hg, FeS and CB showed the best immobilizing ability. Kaolinite presented the weakest removal performance towards all of these five metals, and lack of the presence of kaolinite did not show significant influence on the hydraulic stability of mixed caps.

Based on the lab-scale microcosm experiments, the mixing ratio of 10% kaolinite + 2.5% FeS + 2.5% CB + 35% HAP + 50% OSP (i.e., Column 5) had the most prominent effect to immobilize the five metals present in the test sediment. Although Column 4 (the lowest cost one) with 80% OSP showed unsuccessful results for reducing Ni and MeHg, it performed well in inhibiting the release of the other metals. When considering cost effectiveness and environmental impact, using caps mainly composed of OSP can yet be a good choice. But when applied to real sites, decisions should be made after comprehensive evaluations based on actual conditions.

This study helps to construct guidelines of using mixed materials mostly prepared from recycled materials to remediate multi-contaminated sediments and provide some references for active capping application and development. Results from this study would also be helpful in reducing the human health and ecological risks by reducing the potential toxic metal release from sediment to overlying water.

Supplementary Materials: The following are available online at http://www.mdpi.com/2073-4441/12/7/1886/s1, Figure S1: title, Table S1: Oxidation reduction potential (ORP) during operation, Table S2: pH values of microcosm during operation, Table S3: Dissolved oxygen (DO) (mg L−1) of the overlying water in the microcosm, Table S4: Electrical conductivity (EC) of the overlying water in the microcosm, Table S5: THg concentration of the overlying water in the microcosm, Table S6: MeHg concentration of the overlying water in the microcosm, Table S7: Ni concentration of the overlying water in the microcosm, Table S8: Cr concentration of the overlying water in the microcosm, Table S9: Cu concentration of the overlying water in the microcosm, Table S10: Zn concentration of the overlying water in the microcosm.

Author Contributions: Conceptualization, M.-Y.O., Y.T., T.-C.C., and H.-C.H.; methodology, M.-Y.O., Y.T., B.-L.C., and C.C.; formal analysis, M.-Y.O., Y.T., B.-L.C., and Y.-H.C.; investigation, M.-Y.O.; data curation, M.-Y.O. and Y.-H.C.; writing—original draft preparation, M.-Y.O.; writing—review and editing, T.-C.C. and H.-C.H.; visualization, M.-Y.O.; funding acquisition, H.-C.H. All authors have read and agreed to the published version of the manuscript.

Funding: This project was financially supported by the Taiwan Environmental Protection Administration (No. 08BT547001).

Acknowledgments: We are very grateful to all members of the project team and the Groundwater Pollution Remediation Funds of Taiwan Environmental Protection Administration for their funding support.

Conflicts of Interest: The authors declare no conflict of interest.

References

1. Bouhamed, F.; Elouear, Z.; Bouzid, J.; Ouddane, B. Multi-component adsorption of copper, nickel and zinc from aqueous solutions onto activated carbon prepared from date stones. *Environ. Sci. Pollut. Res.* **2016**, *23*, 15801–15806. [CrossRef]
2. Kutuniva, J.; Mäkinen, J.; Kauppila, T.; Karppinen, A.; Hellsten, S.; Luukkonen, T.; Lassi, U. Geopolymers as active capping materials for in situ remediation of metal (loid)-contaminated lake sediments. *J. Environ. Chem. Eng.* **2019**, *7*, 102852. [CrossRef]
3. Zhang, C.; Zhu, M.Y.; Zeng, G.M.; Yu, Z.G.; Cui, F.; Yang, Z.Z.; Shen, L.Q. Active capping technology: A new environmental remediation of contaminated sediment. *Environ. Sci. Pollut. Res.* **2016**, *23*, 4370–4386. [CrossRef] [PubMed]
4. Race, M.; Nabelkova, J.; Fabbricino, M.; Pirozzi, F.; Raia, P. Analysis of heavy metal sources for urban creeks in the Czech Republic. *Water Air Soil Pollut.* **2015**, *226*, 322. [CrossRef]
5. Salomons, W.; De Rooij, N.M.; Kerdijk, H.; Bril, J. Sediments as a source for contaminants? *Hydrobiologia* **1987**, *149*, 13–30. [CrossRef]

6. Zoumis, T.; Schmidt, A.; Grigorova, L.; Calmano, W. Contaminants in sediments: Remobilisation and demobilisation. *Sci. Total Environ.* **2001**, *266*, 195–202. [CrossRef]
7. Hsu, T.C. Experimental assessment of adsorption of Cu^{2+} and Ni^{2+} from aqueous solution by oyster shell powder. *J. Hazard. Mater.* **2009**, *171*, 995–1000. [CrossRef]
8. Jiang, M.Q.; Jin, X.Y.; Lu, X.Q.; Chen, Z.L. Adsorption of Pb (II), Cd (II), Ni (II) and Cu (II) onto natural kaolinite clay. *Desalination* **2010**, *252*, 33–39. [CrossRef]
9. Latifi, Z.; Jalali, M. Measuring and simulating Co (II) sorption on waste calcite, zeolite and kaolinite. *Nat. Resour. Res.* **2019**, 1–15. [CrossRef]
10. Jalees, M.I.; Farooq, M.U.; Basheer, S.; Asghar, S. Removal of Heavy Metals from Drinking Water Using Chikni Mitti (Kaolinite): Isotherm and Kinetics. *Arabian J. Sci. Eng.* **2019**, *44*, 6351–6359. [CrossRef]
11. Alasadi, A.; Khaili, F.; Awwad, A. Adsorption of Cu (II), Ni (II) and Zn (II) ions by nano kaolinite: Thermodynamics and kinetics studies. *Chem. Int.* **2019**, *5*, 258–268.
12. Jeong, H.Y.; Klaue, B.; Blum, J.D.; Hayes, K.F. Sorption of mercuric ion by synthetic nanocrystalline mackinawite (FeS). *Environ. Sci. Technol.* **2007**, *44*, 7699–7705. [CrossRef] [PubMed]
13. Xiong, Z.; He, F.; Zhao, D.; Barnett, M.O. Immobilization of mercury in sediment using stabilized iron sulfide nanoparticles. *Water Res.* **2009**, *43*, 5171–5179. [CrossRef] [PubMed]
14. Feng, Y.; Gong, J.L.; Zeng, G.M.; Niu, Q.Y.; Zhang, H.Y.; Niu, C.G.; Yan, M.; Deng, J.-H. Adsorption of Cd (II) and Zn (II) from aqueous solutions using magnetic hydroxyapatite nanoparticles as adsorbents. *Chem. Eng. J.* **2010**, *162*, 487–494. [CrossRef]
15. Jang, S.H.; Min, B.G.; Jeong, Y.G.; Lyoo, W.S.; Lee, S.C. Removal of lead ions in aqueous solution by hydroxyapatite/polyurethane composite foams. *J. Hazard. Mater.* **2008**, *152*, 1285–1292. [CrossRef]
16. Krestou, A.; Xenidis, A.; Panias, D. Mechanism of aqueous uranium (VI) uptake by hydroxyapatite. *Miner. Eng.* **2004**, *17*, 373–381. [CrossRef]
17. Narwade, V.N.; Khairnar, R.S.; Kokol, V. In situ synthesized hydroxyapatite—Cellulose nanofibrils as biosorbents for heavy metal ions removal. *J. Polym. Environ.* **2018**, *26*, 2130–2141. [CrossRef]
18. Kumar, G.S.; Girija, E.K.; Venkatesh, M.; Karunakaran, G.; Kolesnikov, E.; Kuznetsov, D. One step method to synthesize flower-like hydroxyapatite architecture using mussel shell bio-waste as a calcium source. *Ceram. Int.* **2017**, *43*, 3457–3461. [CrossRef]
19. Wu, Q.; Chen, J.; Clark, M.; Yu, Y. Adsorption of copper to different biogenic oyster shell structures. *Appl. Surf. Sci.* **2014**, *311*, 264–272. [CrossRef]
20. Moon, D.H.; Wazne, M.; Cheong, K.H.; Chang, Y.Y.; Baek, K.; Ok, Y.S.; Park, J.H. Stabilization of As-, Pb-, and Cu-contaminated soil using calcined oyster shells and steel slag. *Environ. Sci. Pollut. Res.* **2015**, *22*, 11162–11169. [CrossRef]
21. Ahmad, M.; Lee, S.S.; Yang, J.E.; Ro, H.M.; Lee, Y.H.; Ok, Y.S. Effects of soil dilution and amendments (mussel shell, cow bone, and biochar) on Pb availability and phytotoxicity in military shooting range soil. *Ecotoxicol. Environ. Saf.* **2012**, *79*, 225–231. [CrossRef] [PubMed]
22. Ahmad, M.; Lee, S.S.; Lim, J.E.; Lee, S.E.; Cho, J.S.; Moon, D.H.; Hashimoto, Y.; Ok, Y.S. Speciation and phytoavailability of lead and antimony in a small arms range soil amended with mussel shell, cow bone and biochar: EXAFS spectroscopy and chemical extractions. *Chemosphere.* **2014**, *95*, 433–441. [CrossRef] [PubMed]
23. EPA Taiwan. *Investigation of Sediment Pollution Sources and Transmission Mode-Taking Key Rivers as Examples*; Taiwan Environmental Protection Administration: Taipei, Taiwan, 2011; EPA-100-GA102-02-A232.
24. ASTM D2765-16. *Standard Test Methods for Determination of Gel Content and Swell Ratio of Crosslinked Ethylene Plastics*; ASTM International: West Conshohocken, PA, USA, 2016; Available online: www.astm.org (accessed on 1 July 2020).
25. Wu, S.C.; Tsou, H.K.; Hsu, H.C.; Hsu, S.K.; Liou, S.P.; Ho, W.F. A hydrothermal synthesis of eggshell and fruit waste extract to produce nanosized hydroxyapatite. *Ceram. Int.* **2013**, *39*, 8183–8188. [CrossRef]
26. Trinkunaite–Felsen, J.; Prichodko, A.; Semasko, M.; Skaudzius, R.; Beganskiene, A.; Kareiva, A. Synthesis and characterization of iron-doped/substituted calcium hydroxyapatite from seashells Macoma balthica (L.). *Adv. Powder Technol.* **2015**, *26*, 1287–1293. [CrossRef]
27. Sobczak–Kupiec, A.; Malina, D.; Kijkowska, R.; Wzorek, Z. Comparative study of hydroxyapatite prepared by the authors with selected commercially available ceramics. *Digest J. Nanomater. Biostructures* **2012**, *7*, 385–391.

28. Lewis, P.A. *Short-Term Methods for Estimating the Chronic Toxicity of Effluents and Receiving Waters to Freshwater Organisms*; EPA, Environmental Monitoring Systems Laboratory: Cincinnati, OH, USA, 1994.
29. Ting, Y.; Chen, C.; Ch'ng, B.L.; Wang, Y.L.; Hsi, H.C. Using raw and sulfur-impregnated activated carbon as active cap for leaching inhibition of mercury and methylmercury from contaminated sediment. *J. Hazard. Mater.* **2018**, *354*, 116–124. [CrossRef]
30. Ting, Y.; Ch'ng, B.L.; Chen, C.; Ou, M.Y.; Cheng, Y.H.; Hsu, C.J.; Hsi, H.C. A simulation study of mercury immobilization in estuary sediment microcosm by activated carbon/clay-based thin-layer capping under artificial flow and turbation. *Sci. Total Environ.* **2020**, *708*, 135068. [CrossRef]
31. Kwapinski, W.; Byrne, C.M.; Kryachko, E.; Wolfram, P.; Adley, C.; Leahy, J.J.; Hayes, M.H.; Novotny, E. Biochar from biomass and waste. *Waste Biomass Valoriz.* **2010**, *1*, 177–189. [CrossRef]
32. Kay, M.I.; Young, R.A.; Posner, A.S. Crystal structure of hydroxyapatite. *Nature* **1964**, *204*, 1050–1052. [CrossRef]
33. Wang, X.; Yu, R.; Wang, P.; Chen, F.; Yu, H. Co-modification of F− and Fe (III) ions as a facile strategy towards effective separation of photogenerated electrons and holes. *Appl. Surf. Sci.* **2015**, *351*, 66–73. [CrossRef]
34. Fuller, C.C.; Bargar, J.R.; Davis, J.A.; Piana, M.J. Mechanisms of uranium interactions with hydroxyapatite: Implications for groundwater remediation. *Environ. Sci. Technol.* **2002**, *36*, 158–165. [CrossRef] [PubMed]
35. Lee, C.H.; Lee, D.K.; Ali, M.A.; Kim, P.J. Effects of oyster shell on soil chemical and biological properties and cabbage productivity as a liming materials. *Waste Manag.* **2008**, *28*, 2702–2708. [CrossRef] [PubMed]
36. Glatstein, D.A.; Francisca, F.M. Influence of pH and ionic strength on Cd, Cu and Pb removal from water by adsorption in Na-bentonite. *Appl. Clay Sci.* **2015**, *118*, 61–67. [CrossRef]
37. Singh, S.P.; Ma, L.Q.; Tack, F.M.; Verloo, M.G. Trace metal leachability of land-Disposed dredged sediments. *J. Environ. Qual.* **2000**, *29*, 1124–1132. [CrossRef]
38. Li, H.; Shi, A.; Li, M.; Zhang, X. Effect of pH, temperature, dissolved oxygen, and flow rate of overlying water on heavy metals release from storm sewer sediments. *J. Chem.* **2013**, 434012. [CrossRef]
39. Zhai, X.; Li, Z.; Huang, B.; Luo, N.; Huang, M.; Zhang, Q.; Zeng, G. Remediation of multiple heavy metal-contaminated soil through the combination of soil washing and in situ immobilization. *Sci. Total Environ.* **2018**, *635*, 92–99. [CrossRef]
40. Calmano, W.; Hong, J.; Förstner, U. Binding and mobilization of heavy metals in contaminated sediments affected by pH and redox potential. *Water Sci. Technol.* **1993**, *28*, 223–235. [CrossRef]
41. Mehrotra, A.S.; Horne, A.J.; Sedlak, D.L. Reduction of net mercury methylation by iron in Desulfobulbus propionicus (1pr3) cultures: Implications for engineered wetlands. *Environ. Sci. Technol.* **2003**, *37*, 3018–3023. [CrossRef]
42. Parks, J.M.; Johs, A.; Podar, M.; Bridou, R.; Hurt, R.A.; Smith, S.D.; Brandt, C.C.; Palumbo, A.V. The genetic basis for bacterial mercury methylation. *Science* **2013**, *339*, 1332–1335. [CrossRef]
43. Bravo, A.G.; Bouchet, S.; Tolu, J.; Björn, E.; Mateos-Rivera, A.; Bertilsson, S. Molecular composition of organic matter controls methylmercury formation in boreal lakes. *Nat. Commun.* **2017**, *8*, 1–9. [CrossRef]

Article

Metallic Pollution and the Use of Antioxidant Enzymes as Biomarkers in *Bellamya unicolor* (Olivier, 1804) (Gastropoda: Bellamyinae)

Mohammed Othman Aljahdali [1] and Abdullahi Bala Alhassan [1,2,*]

[1] Department of Biological Sciences, Faculty of Sciences, King Abdulaziz University, Jeddah 80203, Saudi Arabia; moaljahdali@kau.edu.sa

[2] Department of Biology, Faculty of Life Science, Ahmadu Bello University, Zaria 810001, Nigeria

* Correspondence: balahassan80@gmail.com or aalhassan0021@stu.kau.edu.sa; Tel.: +966-541046505

Received: 26 December 2019; Accepted: 9 January 2020; Published: 10 January 2020

Abstract: Industrial and domestic discharges of effluent is one of the major causes of heavy metal pollution in aquatic ecosystems. Samples of benthic sediment and freshwater mollusc *Bellamya unicolor* were collected from 5 sites in the River Kaduna to determine heavy metal concentration, their ecological risk, and antioxidant enzymes activities in *Bellamya unicolor*. The results revealed the level of pollution based on heavy metal concentrations across the sites in the order S5 > S3 > S4 > S1 > S2. The ecological risk factor (ErF) revealed that Cd made the highest contribution to pollution, recording the highest ErF (2206.41). Moreover, the results of correlation base multivariate analysis showed that urban and industrial waste were the sources of Cu and Pb in the River Kaduna. The significant positive correlation between metal concentration and antioxidants catalase (CAT) and superoxide dismutase (SOD) was established, with maximum activities of antioxidants at site S5. Results from this study have revealed potential ecological risk as a result of heavy metals pollution in the River Kaduna. Hence the need for approaches and policies be put in place to prevent the discharge of untreated industrial and domestic waste into this aquatic ecosystem.

Keywords: heavy metal; pollution; antioxidant; enzyme; biomarkers; ecological risk

---

1. Introduction

Industrialization and urban development are among the major causes of metal pollution in a natural aquatic ecosystem [1]. The persistent nature and bioaccumulation ability of heavy metals in trophic levels make these metals serious pollutants of the aquatic environment [2–4]. The presence of heavy metals in an aquatic ecosystem either through adsorption or co-precipitation can threaten its biodiversity, and the health of humans depending on the resources of that ecosystem [5,6]. However, biological and chemical factors influence the mobility of heavy metals in aquatic environments by desorption from sediments into the surface water [7]. This made the surface water a major intermediate source of metal pollutants in benthic sediments, which is the definitive receptor [8,9].

Heavy metal concentrations of sediments and benthic organisms such as freshwater mollusc, speciation and several analytical techniques are often used in the evaluation of probable ecological risks in benthic sediments and their effect on the biota [10,11]. Several environmental factors and pollution indices such as geoaccumulation index (I_{geo}), ecological risk factors (ErF) and sediment quality guidelines (SQGs) have been used by several authors. It was used to measure the level or degree of pollution caused by metal and the ecological risk posed by metals in benthic sediments [12]. Although chemical speciation is ignored, it can be considered as subjective when evaluation of the level of pollution is computed based on I_{geo} and ErF. This is done because of the efficiency of these indices in utilizing both the concentration and toxic effect of metals to draw valuable conclusions on risk assessments [13,14].

The discharged industrial and domestic effluent containing toxic metals in aquatic ecosystems is a major concern to the survival of aquatic biota. Freshwater molluscs are not an exception. Although they have devised a means of bioaccumulating the metals even at high levels, due to their high tolerant ability to metal concentrations [15]. This ability makes them a good bio-monitor and bio-indicators of metal pollution in aquatic ecosystems [15,16]. The need to involve freshwater mollusc in ecological risk assessment studies does not stop at their ability to accumulate metal pollutants in high concentration. They also form an important link in the metal cycle for aquatic ecosystems and occupy a trophic level in the aquatic food chain [17,18].

Heavy metal concentrations in high levels can lead to the generation of reactive oxygen species such as H_2O_2, OH, RO_2 among others [19,20]. The manifestation of oxidative stress as a result of oxidative damage to proteins, nucleic acids, and lipids in aquatic organisms exposed to metal-polluted environments is triggered by reactive oxygen species (ROS). This happens when the process of detoxifying metals by organisms has to do with redox cycling reactions [20]. However, another important biomarker of oxidative stress known as defensive antioxidant enzymes is found in organisms [20,21], to scavenge ROS produced in organisms because of metal pollution and oxidative stress. Therefore an increase in ROS formation as a result of an increase in metal pollution triggers an increase in the production of antioxidant enzymes [22,23]. This phenomenon leads to the utilization of antioxidants as biomarkers of environmental pollution, which is a source or primary cause of oxidative stress in organisms [23].

In this study, five sites were selected to evaluate the risk assessment of heavy metals in the River Kaduna. Hence, to achieve the aim of this study, we determined concentrations of heavy metals in sediments and the freshwater mollusc *Bellamya unicolor*, pollution indices and antioxidant enzyme activities in *Bellamya unicolor* across the five sites.

2. Materials and Methods

2.1. Study Area

The study area is located in Kaduna state (Lat. 10.52° N and Lat. 12° N, Long. 7.44° E and Long. 9° E). Kaduna state is located in the northern guinea savannah zone of Nigeria as one of the most developed industrial cities (Figure 1). The textile industry, flour mill, fertilizer, plastic, agrochemical, brewery, and bottling companies are some of the major industries sited in Kaduna state. The climatic conditions like most of Nigeria are characterized by the dry and wet season. The wet season commences either in April or May and ends in October of the same year while the dry season begins towards the end of October and ends in March the following year. Annual rainfall on average is between 1450 to 2000 mm with a temperature regime average of 25 to 43 °C, and relative humidity estimated to be between 20 and 40% in January and 60 to 80% in July. The solar radiation of the Kaduna state was 25.0 Wm^{-2} day^{-1} [24].

The River Kaduna is one of the major rivers in Nigeria. It stretches southwest and south course before completing a flow of 550 KM into River Niger. During its course through Kaduna city, it stretches along the southern part of the state through Kakuri where it receives industrial and domestic wastewater. It serves as a source of domestic water supply and irrigation farming for Kaduna urban settlements and its various industries [25].

2.2. Collection of Samples

Surface sediments (0–10 cm) were collected along a stretch of River Kaduna from five sites base on the type of anthropogenic activities, stages for the stretch of the river either downstream, upper or in between, and history of sediment pollution. Coordinates of the sites selected were determined and recorded using a T10 handheld Global Positioning System (GPS) receiver. At each site, five grab hauls were sampled using an Ekman's grab. The samples were sorted in the field to separate *Bellamya unicolor* from the sediment samples. *Bellamya unicolor* were identified as described by Brown and Kristensen [26]. After sorting, the samples were placed inside zip lock bags and stored in the icebox to be transported to the laboratory for further analysis.

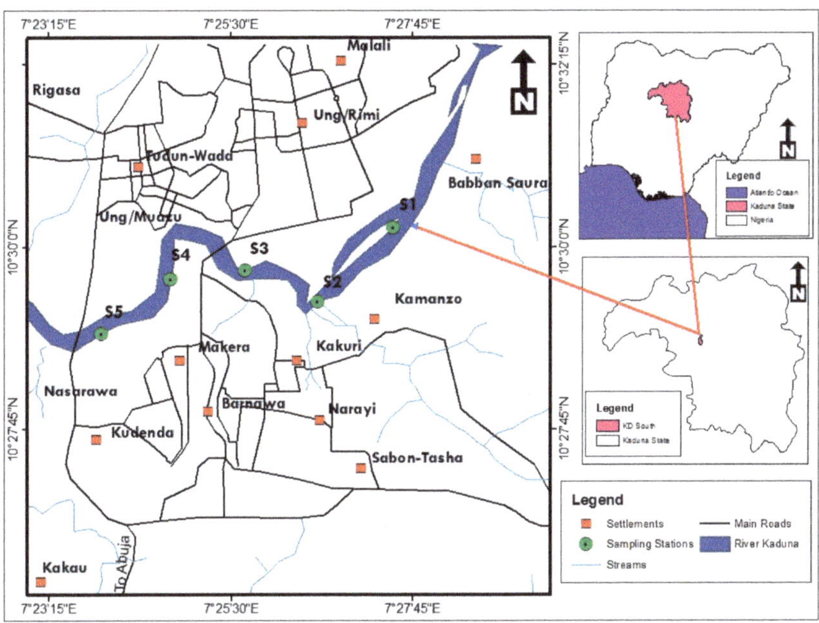

Figure 1. Map of River Kaduna showing sampling sites.

2.3. *Treatment of Samples and Determination of Heavy Metals*

Sediment samples were freeze-dried and grounded while samples of *Bellamya unicolor* were treated and prepared for heavy metal determination and antioxidant enzyme activities assays; 0.4 g of grounded freeze-dried samples was digested in 8 mL of 1:1 HNO_3:HCl and kept for 1 h in Anton-Paar PE Multiwave 3000 (microwave oven) set at 200 °C [1,27].

For the freshwater mollusc *Bellamya unicolor* samples, 0.3 g of the grounded samples were digested chemically in 3:1 HNO_3:H_2O_2 and maintained for 45 min at 180 °C. The volume was then made up to 50 mL using ultra water. The solutions were left to stand after the mixture until the following day. The solute from the digested solution of sediment and *Bellamya unicolor* tissue was then transferred into polypropylene vials after filtration, stored between 4–5 °C and later analyzed for heavy metals using inductively coupled plasma optical emission spectrometry (ICP-OES, Thermo, Waltham, MA, USA).

2.4. *Assay of Antioxidant Enzymes Activities*

Bellamya unicolor tissues were separated from the shell and washed with deionized water; 0.25 g of the tissue sample was then homogenized using 0.01 (M) chilled phosphate buffer at pH 7 with ice-cold mortar and pestle. The homogenate was later centrifuged for 25 min at 4 °C and 14,000 rpm [1,28]. The supernatant from the centrifuged samples was used for the measurement of antioxidant enzyme activities using an LT-291 Single Beam ultraviolet-visible (UV-VIS) spectrophotometer.

2.4.1. Assay of Catalase (CAT)

Catalase (CAT) activities in the supernatant of initially centrifuged samples of *Bellamya unicolor* were assayed as described by Chance and Maehly [29]. CAT activities were estimated spectrophotometrically at 230 nm. The enzyme extract prepared initially was used to measure the activity of catalase. Specific activity was expressed as Ug^{-1} protein. 1 IU = change in absorbance/min/extinction coefficient (0.021).

2.4.2. Assay of Superoxide Dismutase (SOD)

The activity of superoxide dismutase (SOD) was determined using the method of Kakkar et al. [30]. The initial supernatant prepared was purified by precipitating the protein using 90% ammonium sulphate before an assay of enzyme activities. The fraction was then dialysed against 0.0025 M tris-HCl buffer (pH 7.4) and the supernatant was used as the source of enzyme. The content of the assay solution was made up of 1.2 mL of sodium pyrophosphate buffer, 0.1 mL of phenazine methosulphate (PMS), 0.3 mL of nitro-blue tetrazolium (NBT), 1.3 mL of distilled water and 0.1 mL of the enzyme source. The supernatant was kept inside the vials at 30 °C for 90 s and the reaction was stopped by adding 1 mL of glacial acetic acid. 4.0 mL of n-butanol was added to the reacting mixture and mixed properly, then allow to stand for 10 min and the upper portion of the butanol layer was decanted. Absorbance was measured at 560 nm against a blank of n-butanol. A system that lacks enzyme was used as the control and one unit of activities of the enzyme was define as its concentration needed to inhibit 50% production of chromogen per minute. Specific activity was expressed as Ug^{-1} protein. 1 IU = change in absorbance/min/extinction coefficient (0.021).

2.5. Sediment Contamination and Ecological Risk Assessment

The geo-accumulation index (I_{geo}) provides a good explanation of the contamination or pollution status of sediment and thereby gives a better understanding of the possible threat of metal pollutants to natural ecosystem settings [31]. For that reason, I_{geo} was applied in this study as a quantitative indicator, putting into consideration the classifications that evaluate the level of pollution as described by Müller [32]. This index is expressed mathematically as:

$$I_{geo} = log_2[\frac{C_n}{1.5 \times B_n}] \tag{1}$$

where C_n and B_n are the metal concentration in the sediment and geochemical background value of the element (n).

The levels of metal contamination were classified into seven levels: uncontaminated (<0), uncontaminated to moderately contaminated (0–1), moderately contaminated (1–2), moderately to strongly contaminated (2–3), strongly contaminated (3–4), strongly to extremely contaminated (4–5) and extremely contaminated (>5) [32].

In aquatic ecosystems generally, sediment quality guidelines (SQGs) are utilized to assess the potential risk posed by pollutants to the natural ecosystem [33]. Sub variables of SQGs, which are threshold effect limit (TEL) and probable effect limit (PEL), were used for comparison with metal concentrations in sediment to conclude possible potential ecological risk. The concentrations of metals less than the threshold effect limit (TEL) denote a minimal effect, below which there is no expectation for adverse effects. However, metal concentrations at or greater than the probable effect limit (PEL) denote likely frequent occurrence of adverse biological effects (Table 1).

The ecological risk factor (ErF) was used to further assess the status of metal pollution in sediment and its possible toxicological effect with a comprehensive evaluation of ecological risk constituted by metal contaminants. The basis for the classification of ErF values for metal pollutants was as described by Hakanson [34]. For a particular metal, ErF was expressed as:

$$\text{ErF} = Tr^i \times CF^i \tag{2}$$

where Tr^i is the toxic response factor of a given metal (i) (Pb = 5, Zn = 1, Cu = 5 Cd = 30, Cr = 2, Ni = 5, Mn = 1) and CF^i is the contamination factor of metal (i).

The classification of ErF is in five classes according to Hakanson [34]: low risk (ErF < 30), moderate risk (30 ≤ ErF < 60), considerable risk (60 ≤ ErF < 120), high risk (120 ≤ ErF < 240) and very high risk (ErF ≥ 240).

2.6. Data Analysis

SPSS 22.0 and Minitab version 17.0 statistics software were used for data analysis. At 95% confidence interval ($p < 0.05$); analysis of variance (ANOVA) was used to test for significance in mean concentrations of metals across sites both in sediment and the freshwater mollusc *Bellayma unicolor*. Pearson correlation, principal component analysis (PCA) and cluster analysis were used to determine the sources of metal pollutants, the significant relationship between the metals, the relationship between the sites in terms of heavy metal concentration, and the influence of metal concentrations on antioxidant enzymes activities in *Bellamya unicolor*.

3. Results

3.1. Concentrations of Heavy Metals in Sediment and Freshwater Snail

The summary and some important statistics on heavy metal concentrations in sediment and freshwater mollusc from the River Kaduna used sediment quality guidelines (SQGs) to draw some important conclusions in our study which are presented in Table 1. The concentrations of the heavy metals were compared across the five sites studied both in sediment and freshwater mollusc. Site S5 which is the upstream recorded higher concentrations of heavy metals while site (S1) which the upstream recorded the least concentrations of the nine heavy metals studied.

The ranges for heavy metal concentrations in sediments were 34.54–165.32 mg/kg, 6.36–78.98 mg/kg, 1.63–59.01 mg/kg, 4.50–61.48 mg/kg, 0.96–58.84 mg/kg, 16.09–79.20 mg/kg, 39.43–96.08 mg/kg, 2.64–60.08 mg/kg and 11.42–62.61 mg/kg for Fe, Mn, Cu, Zn, Cd, Cr, Pb, Ni and Co respectively. Heavy metal concentrations in freshwater mollusc were lower than the concentrations in the sediment which ranged from Fe: 6.61–33.30, Mn: 1.34–22.75, Cu: 0.54–16.51, Zn: 1.74–18.63, Cd: 0.92–17.52, Cr: 4.26–17.24, Pb: 6.24–24.20, Ni: 0.63–14.17 and Co: 1.6–21.46.

A comparison was made between heavy metal concentrations in benthic sediment samples with TEL and PEL (Table 2). The result of the comparisons revealed that Cu, Cd, Cr, Pb, and Ni fell between TEL and PEL for 46.4, 20.8, 57.4, 100 and 61.2% of the samples respectively. However Cd again also had a value above PEL for 79.2% of the samples.

3.2. Ecological Risk Assessment of Heavy Metals

3.2.1. Geo-Accumulation Index (I_{geo})

In this study, the values of I_{geo} recorded for heavy metal concentrations of sediment samples from the River Kaduna are presented in Table 3. Based on the classification by Muller [32], the recorded values of I_{geo} for Cd at sites S1 (−2.39) and S2 (−2.91) fell into class zero, which means the sediments in S1 and S2 were unpolluted by Cd. In contrast, all the sites were found to be extremely polluted with the other heavy metals with $I_{geo} > 5$ falling in class 6. However, site S4 I_{geo} (0.33) for Cd fell in class 1, and S3 (3.53) and S5 (3.56) fell in class 4 indicating unpolluted to moderate pollution, and strongly polluted respectively.

Table 1. Consensus-based sediment quality guideline values (mg/kg) for heavy metals in sediment and averages for metals concentration in sediments and *Bellamya unicolor* sampled from the River Kaduna.

	Fe	Mn	Cu	Zn	Cd	Cr	Pb	Ni	Co
CSd	90.8225	42.2875	31.285	33.5475	29.4975	46.67625	67.66875	31.6575	34.6275
Range	34.54–165.32	6.36–78.98	1.63–59.01	4.50–61.48	0.96–58.84	16.09–79.20	39.43–96.08	2.64–60.08	11.42–62.61
CSn	20.8825	9.0625	7.295	7.555	7.4225	10.3725	15.0375	7.27	7.435
Range	6.61–33.30	1.34–22.75	0.54–16.51	1.74–18.63	0.92–17.52	4.26–17.24	6.24–24.20	0.63–14.17	1.6–21.46
TEL	NA	NA	35.7	123	0.596	37.3	35	18	NA
PEL	NA	NA	197	315	3.53	90	91.3	36	NA

Note: CSd = Concentration of heavy metals in sediment. CSn = Concentration of heavy metals in freshwater snail: *Bellamya unicolor*. TEL = Threshold effect level. PEL = Probable effect level, NA = Not available.

Table 2. Comparison between sediment quality guidelines (SQGs) and concentration of heavy metals in sediment of the River Kaduna.

SQGs	Fe	Mn	Cu	Zn	Cd	Cr	Pb	Ni	Co
% of samples < TEL	NA	NA	53.6	100	0	42.6	0	38.8	NA
% of samples btw TEL-PEL	NA	NA	46.4	0	20.8	57.4	100	61.2	NA
% of samples > PEL	NA	NA	0	0	79.2	0	0	0	NA

Note: TEL = Threshold effect level. PEL = Probable effect level, NA = Not available.

Table 3. Geoaccumulation (I_{geo}) values of heavy metals in sediments and their class.

	I_{geo}							
Sites	**Mn**	**Cu**	**Zn**	**Cd**	**Cr**	**Pb**	**Ni**	**Co**
S1	11.81	5.61	8.15	−2.39	9.89	9.00	6.42	5.78
S2	12.47	7.04	8.40	−2.91	10.31	9.03	7.51	7.13
S3	15.34	10.79	11.96	3.53	12.09	10.31	11.40	9.62
S4	13.68	8.18	10.07	0.33	10.96	9.72	9.24	7.84
S5	15.44	10.79	11.93	3.56	12.21	10.32	11.41	9.63
Mean	13.75	8.48	10.10	0.42	11.09	9.68	9.19	8.00
Level	6	6	6	1	6	6	6	6

Note: S = Site.

3.2.2. Ecological Risk Factor (ErF)

For ErF results of heavy metal concentrations in the River Kaduna, the level of pollution was in the order Cd > Co > Ni > Pb > Cr > Cu > Zn > Mn (Table 4). Compared to other heavy metals, Cd ErF revealed a high metal pollution level especially at site S5 (ErF = 2206.41). However, the values of ErF for Mn, Cu, Zn, Cr, Pb, Ni, and Co revealed low potential ecological risk or pollution. But in total, the multi-metal ErF as reflected for variation in sites classified the sediments into low pollution to very high pollution in the sequence S5 > S3 > S4 > S1 > S2. Sites S3, S4, and S5 were classified as having very high ecological risk with ErF ≥ 320, S2 as a low potential ecological risk with ErF <40, and S1 as a moderate potential ecological risk with ErF falling between 40–79.9.

Table 4. Enrichment factor (ErF) values of heavy metals in sediment samples from the River Kaduna.

	(Individual Metal)								**(Multi-Metal) ErF**
Sites	**Mn**	**Cu**	**Zn**	**Cd**	**Cr**	**Pb**	**Ni**	**Co**	**ErF**
S1	0.01	0.02	0.03	35.86	0.32	2.27	0.38	1.32	40.21
S2	0.02	0.07	0.04	25.31	0.43	2.32	0.58	3.26	32.03
S3	0.16	0.62	0.44	2160.00	1.45	5.60	8.51	17.74	2194.52
S4	0.09	0.32	0.24	1103.20	0.88	4.07	4.56	9.26	1122.61
S5	0.17	0.62	0.45	2206.41	1.58	5.65	8.58	17.89	2241.35
Mean	0.09	0.33	0.24	1106.16	0.93	3.98	4.52	9.89	1126.14

Note: S = Site.

3.3. *Identification of Pollution Sources*

Pearson correlation analysis (Table 5) revealed significant relationship between the metals except Fe–Mn (0.584), Fe–Cr (0.632), Mn–Cu (0.567), Cu–Zn (0.619), Mn–Pb (0.592), Cu–Cr (0.555), Cu–Ni (0.622), Cr–Pb (0.572) and Pb–Co (0.608). However, Cd concentration at $p < 0.05$ and $p < 0.01$ was significantly correlated with all the other metals concentrations.

Table 5. Pearson correlation analysis among heavy metals and antioxidant enzymes.

	Fe	**Mn**	**Cu**	**Zn**	**Cd**	**Cr**	**Pb**	**Ni**	**Co**	**SOD**	**CAT**
Fe	1										
Mn	0.584	1									
Cu	0.777 **	0.567	1								
Zn	0.671 *	0.984 **	0.619	1							
Cd	0.728 *	0.971 **	0.660 *	0.985 **	1						
Cr	0.632	0.988 **	0.555	0.993 **	0.981 **	1					
Pb	0.765 **	0.592	0.994 **	0.638 *	0.671 *	0.572	1				
Ni	0.672 *	0.978 **	0.622	0.993 **	0.988 **	0.992 **	0.634 *	1			
Co	0.640 *	0.982 **	0.585	0.993 **	0.983 **	0.987 **	0.608	0.986 **	1		
SOD	0.809 **	0.822 **	0.845 **	0.827 **	0.877 **	0.809 **	0.860 **	0.832 **	0.822 **	1	
CAT	0.567	0.779 **	0.758 *	0.765 **	0.769 **	0.735 *	0.801 **	0.759 *	0.777 **	0.904 **	1

Note: ** Correlation is significant at the 0.01 level (2-tailed), * Correlation is significant at the 0.05 level (2-tailed), SOD = Superoxide dismutase, CAT = Catalase.

The similarity in terms of the pattern of distribution of heavy metals was established through the utilisation of principal component analysis (PCA); PCA pulls the data together into a form that can be managed easily and takes out a small number of latent factors to analyse the relationship between the variables observed [35,36]. PCA was applied here primarily to evaluate the source origin of heavy metals in the region.

The results for PCA are presented in Table 6. The high eigenvalue is an indication of pattern and to what extent the data is spread and this leads to consideration of high eigenvalue as the principal component. The components/factors associated with the results of PCA for heavy metals in sediments and freshwater mollusc had a total variation of 92.5%. Component 1 (PC1) recorded the highest eigenvalue of 13.82, accounting for 76.80% of the total variation, and dominate the other components with more significant variation and strong positive loadings (>0.90) of Mn, Cu, Cd, Cr, and Pb in the sediment. Component 2 (PC2) accounted for 8.81% of total variation and had moderate positive loadings (>0.50) of Cu and Pb in the sediment. Component 3 (PC3) accounted for 6.89% of the total variation having moderate positive loadings (>0.50) on Co.

Table 6. Principal component analysis (PCA) loadings for heavy metals in sediment (Sd) and *Bellamya unicolor* (S).

	Component		
Heavy Metal	**PC1**	**PC2**	**PC3**
Fe_S	0.74	0.40	0.03
Mn_S	0.97	−0.20	−0.02
Cu_S	0.73	−0.23	0.01
Zn_S	0.98	−0.17	−0.02
Cd_S	0.99	−0.10	0.00
Cr_S	0.97	−0.23	−0.02
Pb_S	0.75	−0.17	0.05
Ni_S	0.98	−0.16	0.03
Co_S	0.97	−0.22	0.30
Fe_Sd	0.85	0.26	0.21
Mn_Sd	**0.97**	−0.18	0.07
Cu_Sd	**0.97**	**0.63**	−0.01
Zn_Sd	0.68	0.36	−0.05
Cd_Sd	**0.98**	−0.18	−0.05
Cr_Sd	**0.98**	−0.16	0.01
Pb_Sd	**0.97**	**0.60**	−0.09
Ni_Sd	0.67	0.33	−0.52
Co_Sd	0.24	0.07	**0.95**
Eigenvalue	13.82	1.59	1.24
Total variance (%)	76.80	8.81	6.89

Note: Method of extraction: principal component analysis. Rotation method. Bold numbers indicate a strong loading value (>0.9); moderate loading value (>0.5).

Cluster analysis reveals the relationship between sampling sites (Figure 2a) based on heavy metals concentrations and also the relationship between the metals (Figure 2b). For the relationship between the sites, three (3) clusters were formed at a similarity level of 89.09. The first cluster presented only site S1 and the second cluster had only site 2 while the third cluster grouped sites S3, S4, and S5. Four (4) clusters were formed for the relationship between the metals at a similarity level of 99.57. The clusters were Cluster 1 (Fe), Cluster 2 (Mn, Cr, Cu, Cd, and Pb), Cluster 3 (Zn and Ni) and Cluster 4 (Co).

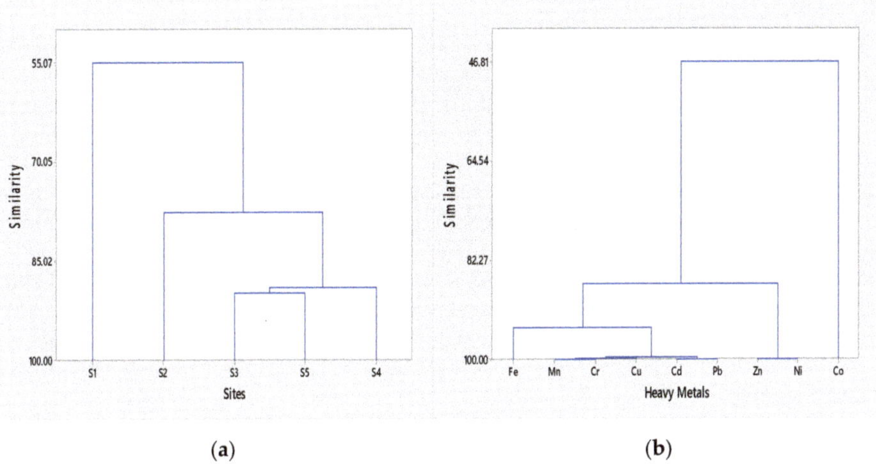

(a) (b)

Figure 2. Cluster analysis for the relationship between (**a**) sites; (**b**) heavy metals.

3.4. Antioxidant Enzyme Activities

Antioxidant enzymes such as CAT and SOD can be used to monitor changes in an environment through the measurement of their activities in an organism present in that same environment. The activities of CAT and SOD in this study are shown in Figure 3a,b respectively. Significant variation existed at $p < 0.05$ across the sites for the two antioxidants with sites S5 recording the highest activities for both CAT and SOD. However, CAT activities ranged from 6.67–72.46 Ug^{-1} protein and SOD 2.70–12.49 Ug^{-1} protein. The sequence for activities of CAT and SOD across the sites; S5 > S3 > S4 > S1 > S2 corresponded with that of heavy metals concentrations, I_{geo} and ErF.

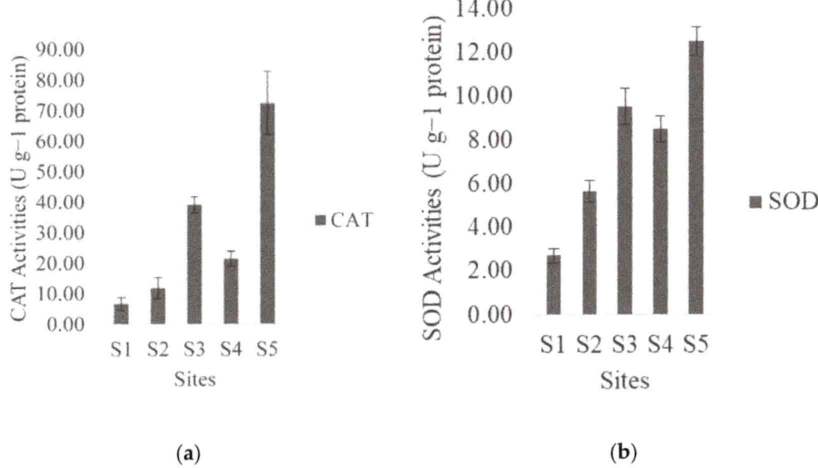

(a) (b)

Figure 3. Antioxidant enzymes activities in *Bellamya unicolor* across the sites for (**a**) catalase (CAT); (**b**) superoxide dismutase (SOD).

A significant positive correlation ($p < 0.01$) and ($p < 0.05$) existed between CAT, SOD and all the heavy metals except Fe (0.567) which did not correlate with CAT significantly.

4. Discussion

The high concentrations of heavy metals recorded in this study downstream (S5) and the significant variations across the sites both in sediment and freshwater molluscs can be as a result of anthropogenic activities from the nearby settlements and the industries sited close to the aquatic ecosystem under study [37,38]. However, even though reduced to non-anthropogenic activities were seen or recorded at site S1 which is located upstream, concentrations of heavy metals were still recorded. This can be due to the atmospheric deposition of heavy metals and the effect of runoff that ends up in the river [17,39]. The deposition of metals from the atmosphere, industries located at the catchment, effluents of agrochemical sources and domestic sewage discharge forms part of the input for metal pollutants of anthropogenic sources. Their effect is reflected in natural aquatic ecosystem sediments with negative consequences to benthic organisms and the general wellbeing of the ecosystem [40,41].

Several authors [4,6,14] report similar findings in ecological assessment of the aquatic ecosystem having variations in different sites, with maximum concentration at sites with industrial plants and more anthropogenic activities.

The concentrations of heavy metals such as Cu, Cd, Cr, Pb and Ni that fall between TEC and PEC values for 46.4, 20.8, 57.4, 100 and 61.2% respectively from sediment quality guidelines of the freshwater ecosystem may be an indication of occasionally serious negative effects of this metals upon the ecosystem under study [4]. However, 79.2% of samples with a concentration of Cd greater than PEL reflected the frequent occurrence of biological effect [42]. Comparison is made between metal concentrations with TEC and PEC values to identify if heavy metals present in the sediment has the possibility of threatening the aquatic life [14,42].

This site has industries sited close to it and receives effluents from the industries. Hence, serious attention might be required at this site due to high concentrations of metal pollutants in comparison with the other sites. Pollution as a result of industrial effluent and domestic waste discharge contributes to the rise in concentrations of metal pollutants leading to contamination of the ecosystem [4,43].

The I_{geo} of heavy metals in sediments were generally higher in site S5 and lower in sites S1 and S2. I_{geo} scoring of S1 and S2 into class zero [32] for Cd indicate no contamination with Cd. Cd is well known as one of the major pollutants with industrial effluents and domestic sewage as its source [43]. However, extreme pollution of the sites with heavy metals especially S5 with $I_{geo} > 5$ having a score of 6 reflects on the source variations of the metal pollutants and the nature of catchments. Site S5 is downstream where the other stretch of the river ends up, receiving some of its contents apart from the effluents received from the nearby industrial plants. Downstream has been reported in several studies to have more pollutants affecting the natural wellbeing of that particular ecosystem setting [44–46].

Considerable ecological risk exhibited by the metals across the sites is linked with the level of geoaccumulation recorded. The ecological risk was consistent with I_{geo} for both metals and site variation as revealed by multi-metal ErF. Cd posed a high ecological risk in River Kaduna with an expectation of an adverse effect expected to occur suggesting that they are present in high concentrations, especially at site S5 with the maximum ErF value. This major ecological risk in surface sediment of the River Kaduna by Cd must have been seriously influenced by anthropogenic activities, leading to a high toxic response factor [6]. The low and moderate potential ecological risk in sites S1 and S2 may be a result of metals present in residual forms and minimal concentrations [31,47,48].

Source and migration of metals may be reflected by correlation analysis and PCA [49,50]. Non-significant correlation between Fe–Mn, Fe–Cr, Mn–Cu, Cu–Zn, Mn–Pb, Cu–Cr, Cu–Ni, Cr–Pb, and Pb–Co indicates different factors controlling the availability and concentrations of these metals [4,9]. This implies the metals might have originated from different sources. This was also revealed in the PCA; the groupings of metals, and component 1 and 2 accounting for 76.80% and 8.81%, respectively, with strong positive loading of Mn, Cu, Cd, Cr and Pb in the sediment and moderate positive loading of Cu and Pb in the sediment. This implies urban and industrial waste to be the source of Mn, Cu, Cd, Cr and Pb [51,52].

Similarity and dissimilarity in groups are effectively represented using cluster analysis. Three cluster formations for sites under study revealed a close relationship between these sites. Grouping of S3, S4, and S5 together at a similarity level of 89.09 indicates possible similarities in human activities, site morphology and the pristine background types of sources [53]. The significant relationship revealed by cluster analysis at 99.57 similarity level for the relationship between the metals, grouping Mn, Cr, Cu, Cd, and Pb together makes it possible that they have the same natural and anthropogenic sources [54,55]. This is also a link to the multiple effects of these metals on benthic organisms. For example, *Bellamya unicolor* collected during our sampling for this study. Our result agrees with the finding of Peijnenburg et al. [56]; Li et al. [55]; Chung et al. [54]. These authors assess and monitor heavy metal sources and their risk in different aquatic environments.

CAT and SOD activities were used in this study as biomarkers to measure the extent of stress in the River Kaduna as reflected in *Bellamya unicolor* sampled from the same environment at different sites. CAT and SOD are a type of antioxidant enzyme that scavenges reactive oxygen species (ROS) (H_2O_2, OH, O_2, etc.) produced by organisms as a result of oxidative stress caused as a result of unfavorable environmental conditions such as metal pollution [1]. Significant variation in CAT and SOD activities across the sites with high activities in site S5 may be a result of different concentrations of ROS production at the sites with high concentrations at site S5 due to high metal pollution [57]. The scavenging ability of antioxidants increases through their activities in response to increased ROS to reduce or prevent membrane lipid peroxidation by ROS and to improve membrane stability of the cell [58,59].

The significant positive correlation between the antioxidants and the metals showed that the organism *Bellamya unicolor* is under stress as a result of an increase in metal contamination. Bakshi et al. [1] report similar findings for the biological response of an aquatic organism to metal contamination through antioxidant activities. They revealed an increase in antioxidants with a spatial and temporal increase in heavy metal concentrations.

Our results in this study tend to bridge the gap in knowledge for metal pollution risk assessment, and the use of CAT and SOD as primary biomarkers in benthic organisms of a tropical ecosystem. There has so far been no report, to the best of our knowledge, assessing risk assessment using ecological indices, antioxidant enzymes and *Bellamya unicolor* as a biomonitor. Our study will provide data to be used as a baseline for studies in the River Kaduna and other related tropical rivers in the same region.

5. Conclusions

The results from this study provide information on the contamination of heavy metals Fe, Mn, Cu, Zn, Cd, Cr, Pb, Ni, and Co in surface sediment of the River Kaduna and their bioavailability in the freshwater mollusc *Bellamya unicolor* sampled from the same ecosystem.

At a 95% confidence interval, significant variation existed in the concentration of metals in surface sediment and freshwater molluscs across the sampling sites with the site S5 (downstream) having the maximum metal concentrations.

Cd concentration in sediment was greater than PEL, which implies likely frequent occurrence of biological effect by Cd in the River Kaduna with the possibility of threatening the aquatic life. This was reflected in Cd ErF values that are higher than the values of ErF for other metals. I_{geo} and ErF values follow the same sequence with maximum values at site S5 revealing very high ecological risk with $I_{geo} > 5$ and ErF > 320. Mn, Cu, Cd, Cr, and Pb originate mainly from urban and industrial waste as revealed by PC1 and PC2 analysis while S3, S2, and S3 have similar anthropogenic activities and natural metal inputs as shown in cluster analysis. Significant variations existed in CAT and SOD activities in *Bellamya unicolor* across the sites with maximum activities recorded at sites S5 for both antioxidants. This also coincides with the high values of I_{geo} and ErF at site S5. However, there is a significant positive correlation between the antioxidants and metal concentrations.

Important approaches and policies should be put in place to prevent the discharge of untreated industrial and domestic waste into the River Kaduna. The approach should involve the prevention of irrigation farming close to the river and, in that way, non-point sources of pollution can be abated and there can be a decrease in ecological risk associated with metal pollutants.

Author Contributions: M.O.A., and A.B.A. conceived the idea and performed the laboratory analysis. All of the authors contributed to give the manuscript its present shape. All authors have read and agreed to the published version of the manuscript.

Funding: This research was funded by the Deanship of Scientific Research (DSR), King Abdulaziz University, Jeddah, under grant number D-024-130-1441. The authors, therefore, gratefully acknowledge DSR technical and financial support.

Acknowledgments: The authors acknowledge the Deanship of Scientific Research (DSR), King Abdulaziz University, Jeddah for their technical and financial support.

Conflicts of Interest: The authors declare there is no conflict of interest.

References

1. Bakshi, M.; Ghosh, S.; Chakraborty, D.; Hazra, S.; Chaudhuri, P. Assessment of potentially toxic metal (PTM) pollution in mangrove habitats using biochemical markers: A case study on Avicennia officinalis L. in and around Sundarban, India. *Mar. Pollut. Bull.* **2018**, *133*, 157–172. [CrossRef]
2. Bastami, K.D.; Bagheri, H.; Kheirabadi, V.; Zaferani, G.G.; Teymori, M.B.; Hamzehpoor, A.; Soltani, F.; Haghparast, S.; Harami, S.R.M.; Ghorghani, N.F. Distribution and ecological risk assessment of heavy metals in surface sediments along southeast coast of the Caspian Sea. *Mar. Pollut. Bull.* **2014**, *81*, 262–267. [CrossRef] [PubMed]
3. Xiao, R.; Bai, J.; Lu, Q.; Zhao, Q.; Gao, Z.; Wen, X.; Liu, X. Fractionation, transfer, and ecological risks of heavy metals in riparian and ditch wetlands across a 100-year chronosequence of reclamation in an estuary of China. *Sci. Total Environ.* **2015**, *517*, 66–75. [CrossRef] [PubMed]
4. Ke, X.; Gui, S.; Huang, H.; Zhang, H.; Wang, C.; Guo, W. Ecological risk assessment and source identification for heavy metals in surface sediment from the Liaohe River protected area, China. *Chemosphere* **2017**, *175*, 473–481. [CrossRef]
5. Suresh, G.; Ramasamy, V.; Sundarrajan, M.; Paramasivam, K. Spatial and vertical distributions of heavy metals and their potential toxicity levels in various beach sediments from high-background-radiation area, Kerala, India. *Mar. Pollut. Bull.* **2015**, *91*, 389–400. [CrossRef] [PubMed]
6. Xu, M.; Sun, W.; Wang, R. Spatial Distribution and Ecological Risk Assessment of Potentially Harmful Trace Elements in Surface Sediments from Lake Dali, North China. *Water* **2019**, *11*, 2544. [CrossRef]
7. Hill, N.A.; Simpson, S.L.; Johnston, E.L. Beyond the bed: Effects of metal contamination on recruitment to bedded sediments and overlying substrata. *Environ. Pollut.* **2013**, *173*, 182–191. [CrossRef] [PubMed]
8. Bermejo, J.S.; Beltrán, R.; Ariza, J.G. Spatial variations of heavy metals contamination in sediments from Odiel river (Southwest Spain). *Environ. Int.* **2003**, *29*, 69–77. [CrossRef]
9. Kükrer, S.; Şeker, S.; Abacı, Z.T.; Kutlu, B. Ecological risk assessment of heavy metals in surface sediments of northern littoral zone of Lake Çıldır, Ardahan, Turkey. *Environ. Monit. Assess.* **2014**, *186*, 3847–3857. [CrossRef]
10. Yang, Z.; Wang, Y.; Shen, Z.; Niu, J.; Tang, Z. Distribution and speciation of heavy metals in sediments from the mainstream, tributaries, and lakes of the Yangtze River catchment of Wuhan, China. *J. Hazard. Mater.* **2009**, *166*, 1186–1194. [CrossRef]
11. Yu, G.; Liu, Y.; Yu, S.; Wu, S.; Leung, A.; Luo, X.; Xu, B.; Li, H.; Wong, M.H. Inconsistency and comprehensiveness of risk assessments for heavy metals in urban surface sediments. *Chemosphere* **2011**, *85*, 1080–1087. [CrossRef] [PubMed]
12. Zahra, A.; Hashmi, M.Z.; Malik, R.N.; Ahmed, Z. Enrichment and geo-accumulation of heavy metals and risk assessment of sediments of the Kurang Nallah—Feeding tributary of the Rawal Lake Reservoir, Pakistan. *Sci. Total Environ.* **2014**, *470*, 925–933. [CrossRef] [PubMed]
13. Zhu, H.N.; Yuan, X.Z.; Zeng, G.M.; Jiang, M.; Liang, J.; Zhang, C.; Juan, Y.; Huang, H.J.; Liu, Z.F.; Jiang, H.W. Ecological risk assessment of heavy metals in sediments of Xiawan Port based on modified potential ecological risk index. *Trans. Nonferrous Met. Soc. China* **2012**, *22*, 1470–1477. [CrossRef]
14. Rahman, M.S.; Hossain, M.B.; Babu, S.O.F.; Rahman, M.; Ahmed, A.S.; Jolly, Y.; Choudhury, T.; Begum, B.; Kabir, J.; Akter, S. Source of metal contamination in sediment, their ecological risk, and phytoremediation ability of the studied mangrove plants in ship breaking area, Bangladesh. *Mar. Pollut. Bull.* **2019**, *141*, 137–146. [CrossRef]

15. Waykar, B.; Petare, R. Studies on monitoring the heavy metal contents in water, sediment and snail species in Latipada reservoir. *J. Environ. Biol.* **2016**, *37*, 585.
16. Zhou, Q.; Zhang, J.; Fu, J.; Shi, J.; Jiang, G. Biomonitoring: An appealing tool for assessment of metal pollution in the aquatic ecosystem. *Anal. Chim. Acta* **2008**, *606*, 135–150. [CrossRef]
17. Chen, M.; Boyle, E.A.; Switzer, A.D.; Gouramanis, C. A century long sedimentary record of anthropogenic lead (Pb), Pb isotopes and other trace metals in Singapore. *Environ. Pollut.* **2016**, *213*, 446–459. [CrossRef]
18. Shalaby, B.; Samy, Y.M.; Mashaly, A.O.; El Hefnawy, M.A.A. Comparative Geochemical Assessment of Heavy Metal Pollutants among the Mediterranean Deltaic Lakes Sediments (Edku, Burullus and Manzala), Egypt. *Egypt. J. Chem.* **2017**, *60*, 361–378. [CrossRef]
19. Halliwell, B.; Gutteridge, J.M. *Free Radicals in Biology and Medicine*; Oxford University Press: Oxford, UK, 2015.
20. Gawad, S.S.A. Concentrations of heavy metals in water, sediment and mollusc gastropod, Lanistes carinatus from Lake Manzala, Egypt. *Egypt. J. Aquat. Res.* **2018**, *44*, 77–82. [CrossRef]
21. Buhler, D.R.; Williams, D.E. The role of biotransformation in the toxicity of chemicals. *Aquat. Toxicol.* **1988**, *11*, 19–28. [CrossRef]
22. Regoli, F.; Principato, G.; Bertoli, E.; Nigro, M.; Orlando, E. Biochemical characterization of the antioxidant system in the scallop Adamussium colbecki, a sentinel organism for monitoring the Antarctic environment. *Polar Biol.* **1997**, *17*, 251–258. [CrossRef]
23. Siwela, A.H.; Nyathi, C.; Naik, Y.S. A comparison of metal levels and antioxidant enzymes in freshwater snails, Lymnaea natalensis, exposed to sediment and water collected from Wright Dam and Lower Mguza Dam, Bulawayo, Zimbabwe. *Ecotoxicol. Environ. Saf.* **2010**, *73*, 1728–1732. [CrossRef] [PubMed]
24. NIMET. *Nigeria Meteorological Agency*; CRC Press: Kaduna, Nigeria, 2010.
25. KEPA. Kaduna State Environmental Protection Authority, Revised in 1998:1998. Available online: http://www.kepa.org.ng/ (accessed on 31 December 2019).
26. Brown, D.; Kristensen, T. A Field Guide to African Freshwater Snails I. West African species. *Dan. Bilharz. Lab. Charlottenlund* **1993**, *32*.
27. United States Environmental Protection Agency (USEPA). *Method 3051 A. Microwave Assisted Acid Digestion of Sediments, Sludge's, Soils and Oils*; USEPA, U.S. Government Printing Office: Washington, DC, USA, 1997. Available online: http://www.epa.gov/SW-846/pdfs/3051a (accessed on 31 December 2019).
28. Banaee, M.; Sureda, A.; Taheri, S.; Hedayatzadeh, F. Sub-lethal effects of dimethoate alone and in combination with cadmium on biochemical parameters in freshwater snail, Galba truncatula. *Comp. Biochem. Physiol. Part C Toxicol. Pharm.* **2019**, *220*, 62–70. [CrossRef] [PubMed]
29. Chance, B.; Maehly, A. [136] Assay of catalases and peroxidases. *Methods Enzymol.* **1955**, *2*, 764–775.
30. Kakkar, P.; Das, B.; Viswanathan, P. A modified spectrophotometric assay of superoxide dismutase. *Indian J. Biochem. Biophys.* **1984**, *21*, 130–132.
31. Cheng, H.; Li, M.; Zhao, C.; Yang, K.; Li, K.; Peng, M.; Yang, Z.; Liu, F.; Liu, Y.; Bai, R. Concentrations of toxic metals and ecological risk assessment for sediments of major freshwater lakes in China. *J. Geochem. Explor.* **2015**, *157*, 15–26. [CrossRef]
32. Muller, G. Schwermetalle in den sedimenten des Rheins-Veranderungen seit. *Umschav* **1979**, *79*, 133–149.
33. Wang, Y.; Yang, L.; Kong, L.; Liu, E.; Wang, L.; Zhu, J. Spatial distribution, ecological risk assessment and source identification for heavy metals in surface sediments from Dongping Lake, Shandong, East China. *Catena* **2015**, *125*, 200–205. [CrossRef]
34. Hakanson, L. An ecological risk index for aquatic pollution control. A sedimentological approach. *Water Res.* **1980**, *14*, 975–1001. [CrossRef]
35. Loska, K.; Wiechuła, D. Application of principal component analysis for the estimation of source of heavy metal contamination in surface sediments from the Rybnik Reservoir. *Chemosphere* **2003**, *51*, 723–733. [CrossRef]
36. Ma, X.; Zuo, H.; Tian, M.; Zhang, L.; Meng, J.; Zhou, X.; Min, N.; Chang, X.; Liu, Y. Assessment of heavy metals contamination in sediments from three adjacent regions of the Yellow River using metal chemical fractions and multivariate analysis techniques. *Chemosphere* **2016**, *144*, 264–272. [CrossRef] [PubMed]
37. Chai, M.; Shi, F.; Li, R.; Shen, X. Heavy metal contamination and ecological risk in Spartina alterniflora marsh in intertidal sediments of Bohai Bay, China. *Mar. Pollut. Bull.* **2014**, *84*, 115–124. [CrossRef] [PubMed]
38. Cui, J.; Zang, S.; Zhai, D.; Wu, B. Potential ecological risk of heavy metals and metalloid in the sediments of Wuyuer River basin, Heilongjiang Province, China. *Ecotoxicology* **2014**, *23*, 589–600. [CrossRef]

39. Lin, Q.; Liu, E.; Zhang, E.; Nath, B.; Shen, J.; Yuan, H.; Wang, R. Reconstruction of atmospheric trace metals pollution in Southwest China using sediments from a large and deep alpine lake: Historical trends, sources and sediment focusing. *Sci. Total Environ.* **2018**, *613*, 331–341. [CrossRef]
40. Sutherland, R. Bed sediment-associated trace metals in an urban stream, Oahu, Hawaii. *Environ. Geol.* **2000**, *39*, 611–627. [CrossRef]
41. Liu, E.; Birch, G.F.; Shen, J.; Yuan, H.; Zhang, E.; Cao, Y. Comprehensive evaluation of heavy metal contamination in surface and core sediments of Taihu Lake, the third largest freshwater lake in China. *Environ. Earth Sci.* **2012**, *67*, 39–51. [CrossRef]
42. Zhang, R.; Zhou, L.; Zhang, F.; Ding, Y.; Gao, J.; Chen, J.; Yan, H.; Shao, W. Heavy metal pollution and assessment in the tidal flat sediments of Haizhou Bay, China. *Mar. Pollut. Bull.* **2013**, *74*, 403–412. [CrossRef]
43. Zhang, J.; Wang, S.; Xie, Y.; Wang, X.; Sheng, X.; Chen, J. Distribution and pollution character of heavy metals in the surface sediments of Liao River. *Huan Jing Ke Xue Huanjing Kexue* **2008**, *29*, 2413–2418.
44. Mummullage, N.; Wasanthi, S. Source Characterisation of Urban Road Surface Pollutants for Enhanced Water Quality Predictions. Ph.D. Thesis, Queensland University of Technology, Brisbane, Australia, 2015.
45. Brady, J.P.; Kinaev, I.; Goonetilleke, A.; Ayoko, G.A. Comparison of partial extraction reagents for assessing potential bioavailability of heavy metals in sediments. *Mar. Pollut. Bull.* **2016**, *106*, 329–334. [CrossRef]
46. Duodu, G.O.; Goonetilleke, A.; Ayoko, G.A. Potential bioavailability assessment, source apportionment and ecological risk of heavy metals in the sediment of Brisbane River estuary, Australia. *Mar. Pollut. Bull.* **2017**, *117*, 523–531. [CrossRef] [PubMed]
47. Brady, J.P. Heavy Metals in the Sediments of Northern Moreton Bay, Queensland, Australia. Ph.D. Thesis, Queensland University of Technology, Brisbane, Australia, 2015.
48. Liu, R.; Bao, K.; Yao, S.; Yang, F.; Wang, X. Ecological risk assessment and distribution of potentially harmful trace elements in lake sediments of Songnen Plain, NE China. *Ecotoxicol. Environ. Saf.* **2018**, *163*, 117–124. [CrossRef] [PubMed]
49. Wang, Y.; Hu, J.; Xiong, K.; Huang, X.; Duan, S. Distribution of heavy metals in core sediments from Baihua Lake. *Procedia Environ. Sci.* **2012**, *16*, 51–58. [CrossRef]
50. Suresh, G.; Ramasamy, V.; Meenakshisundaram, V.; Venkatachalapathy, R.; Ponnusamy, V. Influence of mineralogical and heavy metal composition on natural radionuclide concentrations in the river sediments. *Appl. Radiat. Isot.* **2011**, *69*, 1466–1474. [CrossRef]
51. Mohiuddin, K.; Otomo, K.; Ogawa, Y.; Shikazono, N. Seasonal and spatial distribution of trace elements in the water and sediments of the Tsurumi River in Japan. *Environ. Monit. Assess.* **2012**, *184*, 265–279. [CrossRef]
52. Chandra, R.; Yadav, S.; Yadav, S. Phytoextraction potential of heavy metals by native wetland plants growing on chlorolignin containing sludge of pulp and paper industry. *Ecol. Eng.* **2017**, *98*, 134–145. [CrossRef]
53. Varol, M.; Şen, B. Assessment of nutrient and heavy metal contamination in surface water and sediments of the upper Tigris River, Turkey. *Catena* **2012**, *92*, 1–10. [CrossRef]
54. Chung, C.Y.; Chen, J.J.; Lee, C.G.; Chiu, C.Y.; Lai, W.L.; Liao, S.W. Integrated estuary management for diffused sediment pollution in Dapeng Bay and neighboring rivers (Taiwan). *Environ. Monit. Assess.* **2011**, *173*, 499–517. [CrossRef]
55. Li, J.; He, M.; Han, W.; Gu, Y. Analysis and assessment on heavy metal sources in the coastal soils developed from alluvial deposits using multivariate statistical methods. *J. Hazard. Mater.* **2009**, *164*, 976–981. [CrossRef]
56. Peijnenburg, W.J.; Zablotskaja, M.; Vijver, M.G. Monitoring metals in terrestrial environments within a bioavailability framework and a focus on soil extraction. *Ecotoxicol. Environ. Saf.* **2007**, *67*, 163–179. [CrossRef]
57. Harish, S.; Murugan, K. Oxidative stress indices in natural populations of Avicennia alba Blume. as biomarker of environmental pollution. *Environ. Res.* **2011**, *111*, 1070–1073. [CrossRef] [PubMed]
58. Shahid, M.; Pourrut, B.; Dumat, C.; Nadeem, M.; Aslam, M.; Pinelli, E. Heavy-metal-induced reactive oxygen species: Phytotoxicity and physicochemical changes in plants. In *Reviews of Environmental Contamination and Toxicology*; Springer: New York, NY, USA, 2014; Volume 232, pp. 1–44.
59. Asaeda, T.; Barnuevo, A. Oxidative stress as an indicator of niche-width preference of mangrove Rhizophora stylosa. *Forest Ecol. Manag.* **2019**, *432*, 73–82. [CrossRef]

Article

Evaluation of the Effect of Gold Mining on the Water Quality in Monterrey, Bolívar (Colombia)

Alison Martín [1,2], Juliana Arias [1], Jennifer López [1], Lorena Santos [1], Camilo Venegas [1], Marcela Duarte [3], Andrés Ortíz-Ardila [4], Nubia de Parra [3], Claudia Campos [1] and Crispín Celis Zambrano [2,*]

1 Department of Microbiology, Laboratorio de Indicadores de Calidad de Agua y Lodos (LIAL), Pontificia Universidad Javeriana, Carrera 7 No. 43–82, Bogotá 110231, Colombia; alison.martin@javeriana.edu.co (A.M.); julichaque@gmail.com (J.A.); jetalomo@gmail.com (J.L.); lorena28sant@gmail.com (L.S.); c.venegas@javeriana.edu.co (C.V.); campos@javeriana.edu.co (C.C.)

2 Department of Chemistry, Pontificia Universidad Javeriana, Carrera 7 No. 43–82, Bogotá 110231, Colombia

3 Independient Investigator, Bogotá 110111, Colombia; duartemarce@yahoo.com (M.D.); nubiapdeparra@gmail.com (N.d.P.)

4 Hydraulic and Environmental Engineering Department, Pontificia Universidad Católica de Chile, Av. Vicuña Mackenna 4860, Macul, Santiago 7820436, Chile; adortiz@uc.cl

* Correspondence: crispin.celis@javeriana.edu.co

Received: 7 August 2020; Accepted: 4 September 2020; Published: 10 September 2020

Abstract: Gold mining uses chemicals that are discharged into rivers without any control when there are no good mining practices, generating environmental and public health problems, especially for downstream inhabitants who use the water for consumption, as is the case in Monterrey township, where the Boque River water is consumed. In this study, we evaluate Boque River water quality analyzing some physicochemical parameters such as pH, heavy metals, Hg, and cyanide; bioassays (*Lactuca sativa*, *Hydra attenuata*, and *Daphnia magna*), mutagenicity (Ames test), and microbiological assays. The results show that some physicochemical parameters exceed permitted concentrations (Hg, Cd, and cyanide). *D. magna* showed sensitivity and *L. sativa* showed inhibition and excessive growth in the analyzed water. Mutagenic values were obtained for all of the sample stations. The presence of bacteria and somatic coliphages in the water show a health risk to inhabitants. In conclusion, the presence of Cd, Hg, and cyanide in the waters for domestic consumption was evidenced in concentrations that can affect the environment and the health of the Monterrey inhabitants. The mutagenic index indicates the possibility of mutations in the population that consumes this type of water. Bioassays stand out as an alert system when concentrations of chemical contaminants cannot be analytically detected.

Keywords: bioassays; gold mining; health risk; mercury; microbiological indicators; mutagenicity; toxicity

1. Introduction

Gold mining in developing countries is the main source of income for 30 million miners globally. About 12% of global gold production is through illegal mining that provides a significant economic benefit to miners but also proves hazardous/harmful for the environment by causing impacts such as water source sedimentation, land cover degradation, deforestation, soil degradation, and chemical contamination with mercury, cyanide, nitric acid, and zinc [1–3]. In Colombia, despite the various alternatives to avoid the use of Hg in gold extraction, the use of the elemental Hg–Au amalgamation method in small-scale artisanal mining areas is extensive [4,5].

Within the gold mining protocols, mercury and cyanide play an important role. These materials are easy to use, available at a low-cost, and easily accessible. However, there is little awareness among the users or villagers about the use risk of cyanide and mercury in the gold extraction process [1,2]. This activity has led to serious pollution of terrestrial and aquatic ecosystems in emerging countries, impacting mining and fishing communities, and also these polluting elements can reach human beings [4–7].

According to the records of the Colombian Mining Association (ACM), gold production increased in 2020, going from 8.9 tons in 2019 to 9.5 tons in the first quarter of 2020, this represents a growth of 7% [8]. On the other hand, for gold extraction, 86% is considered illegal, taking place without a recognized mining title or without being registered. Medium-scale mining constitutes up to 26% and large-scale mining only takes up 2% of the total [9,10]. The population of the Bolívar department is 2,195,811 inhabitants, according to DANE's (National Statistics Administrative Department) projection for 2019 [11]. According to Carranza-Lopez et al. [4] the gold-mining districts (GMDs) at the department of Bolívar have extensive Hg contamination, and this situation requires special attention to reduce environmental and human health problems.

Municipalities of Montecristo, Santa Rosa del Sur, San Martín de Loba, Morales, San Pablo, Barranco de Loba, and Simití that are in Bolívar are where gold mining mainly takes place. Simití is known as the municipality that has the largest gold mining activity within the Bolívar department. It has an estimated population of 10,360 inhabitants in an area of 1345 km^2, the mining activity occurs in the Boque River, which flows in Simití. It starts on Serranía San Lucas, passes through Monterrey district, and flows into Magdalena River [12–14] (Figure 1).

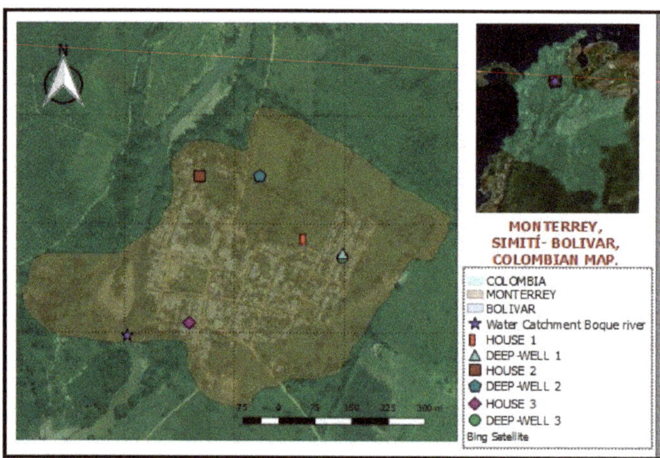

Figure 1. Location of sampling stations in Monterrey, Simití-Bolívar, Colombia. Source: authors.

Gold mining in Middle Magdalena has been carried out through artisanal practices, without considering the implications in the community and ecosystems due to the practice of non-regulated techniques affecting the environment, natural resources as well as health conditions and welfare of the population. Gold mining severely affects water resources, biodiversity, animals, flora, and fauna in its geographical area. In addition, the presence of certain types of mining settlements bring to pass certain types of domestic wastewater discharges without treatment to the Boque River, affecting the quality of the water and the inhabitants downstream [15–19]. The discharge of wastewater into a water body involves a large number and diversity of heavy toxic chemicals, many of which are unknown. These chemicals may react with each other, which can increase the toxicity level, which creates a negative impact on the structure and functioning of the natural ecosystem [4,20].

To determine the effect of gold extraction in the region, the evaluation of physicochemical parameters of the water is required. Nevertheless, the illegal settlements do not have sanitation systems, so microbiological contamination becomes an additional problem. However, even if in some of the above-mentioned situations the parameters could be between the legal requirements, it should be considered that trace heavy elements might have an impact on the population and the ecosystem after long periods of exposure. Thus, it is necessary to test different representatives of the trophic chain to identify the impact of the pollutants through bioassays tests [20–23].

Bioassays are described as alert mechanisms for long-term periods of exposure to chemical pollutants. These are used as indicators of substances that are harmful to living cells and tissues, useful even in the cases where physicochemical parameters fulfill the requirements of water quality [24]. Likewise, this possible bioaccumulation of chemical elements in the trophic chains can generate mutagenicity or toxicity, which is why it is important to be able to establish whether a complex system such as a water sample from a mining region has these undesirable characteristics, which can be detected by the Ames test or bioassays [25–27]. Some Latin American countries have made progress in the application of toxicity tests, while for Colombia, toxicity tests in natural environments are scarce compared to the evaluation of hazardous waste and industrial dumping [28,29]. On the other hand, the Ames test has proven to be effective for the identification of potentially carcinogenic or mutagenic chemicals, achieving its immediate adoption and its requirement by regulatory authorities around the world [30]. In the Ames test, *Salmonella typhimurium* (*S. typhimurium*) is used as an indicator of bacterial mutagenesis as a consequence of exposure to chemical contaminants [25,26].

Taking into consideration that the water of the Boque River is used in human consumption without treatment and it collects chemical pollutants from the mining activity, such as mercury and cyanide, the use of the Ames test in the evaluation of this water will permit the evaluation of its possible carcinogenic or mutagenic effect, making it a relevant issue for the inhabitants of Simití. For this reason, it is necessary to have data on bioassays and Ames test indicators in environmental samples, especially in mining, which has become one of the most important fonts of economic resources in Colombia and at the same time of damage not sufficiently evaluated to date. In order to have a complete evaluation of the water quality in relation to the possible presence of bacteria, viruses, and parasites and the risk to the inhabitants, it is necessary to use indicators of fecal contamination, with the most used indicators being total coliforms and *Escherichia coli* as bacterial indicators and somatic phages as viral indicators, which allow indication of the presence of pathogenic microorganisms in the water.

The aim of this research is to evaluate the impact generated by the exploitation of bad mining practices such as the use of dangerous chemical compounds in gold mining, which are drained into surface waters such as the Boque River in the South of Bolívar, Colombia, as well as the waste generated in the mining settlements. The assessment of the impact on the environment, living organisms, and human health will be done through the detection of heavy metals, microbiological indicators, and bioassays, which through a joint assessment will provide important aspects to protect the health of the inhabitants of Monterrey.

2. Materials and Methods

2.1. Study Area

The Boque River has an approximate area of 876 km^2 and merges into the Magdalena River. The Monterrey district belongs to the Municipality of Simití, Department of Bolívar (Colombia). The inhabitants who live within the Monterey township collect and use the water from the Boque River for different activities, this being the main source of water supply (Figure 1) [27].

2.2. Water Physicochemical Analysis, Heavy Metals, and Cyanide Detection

Some physicochemical parameters were analyzed such as pH (pH/T tester pHep®4, Hanna Instruments, RI, USA) [31], chemical oxygen demand (COD photometer Hanna Instruments, New England, RI, USA) [32], and total solids by the gravimetric method [33]. The detection of heavy metals was performed using a Varian SpectrAA 220 G Atomic Absorption Spectrometer(Varian-Agilent

Inst., Palo Alto, San Francisco, CA, USA), following previous publications: cadmium, chromium, zinc, and nickel [34]; mercury in a direct mercury analyzer (DMA-80, Milestone Inc., Sorisole, Italy) [35]; and cyanide in a portable photometer (Hanna Instruments, RI, USA) [36]. All reactive, analytical standards and reference materials were purchased from Merck (Merck KGaA, Darmstadt, Germany). The results obtained in the samples from Village Gato, Village Tigui, and the water catchment of the Boque River were compared with normative 0631/2015 [37], which establishes the parameters to be monitored and maximum permissible limits in the specific discharges of non-domestic wastewater (precious minerals and gold). While for houses and deep-well underground sites, they were evaluated in compliance with normative 2115/2007 [38], which regulates water for human consumption.

2.3. *Bioassays*

In the bioassays, *Lactuca sativa (L. sativa)* [39] and *Hydra attenuata (H. attenuata)* [40] were used as a biological indicators of water quality. After the follow-up of the results of the two first collections of water samples, a modification of the protocol was performed replacing *Daphnia magna (D. magna)* [41] instead of *H. attenuata* due to no evidence of sensitivity against the possible harmful substances that might be present in the water samples by *H. attenuata*. The effects on organisms can be inhibition, sublethality, and lethality volume/volume (v/v). The water samples were diluted in four different concentrations 25%, 50%, 75%, and 100% (v/v); reconstituted hard water (160–180 mg/L $CaCO_3$) was used as diluent for the *D. magna* and *H. attenuata*; while for *L. sativa*, distilled water was used. The response of *H. attenuata* was read using a binocular stereoscope (Leica). Before taking the readings, the containers were shaken in a circular way to reactivate the movement of *D. magna* and confirm their state. In the case of *L. sativa*, graph paper was used to measure the length of the radicle.

2.3.1. Endpoint and Toxic Response Model

L. sativa, half-maximal inhibitory concentration (IC_{50}): root growth reduction or inhibitory effects on lettuce seed germination and root growth after 5 days. *D. magna*, lethal concentration (LC_{50}): number dead/total number or lethal effects of water were observed after 48 h of exposure, and *H. attenuata*, median effective concentration (EC_{50}): density reduction or lethality test, produced by irreversible morphological changes after 96 h of exposure.

2.3.2. Toxicity

To calculate the lethal concentration (LC_{50}), the half-maximal inhibitory concentration (IC_{50}), and median effective concentration (EC_{50}), the method used was Environmental Protection Agency (EPA) Probit analysis model [42–44]. When results in EC_{50}/LC_{50}/IC_{50} cannot be reported by the statistical program, they are reported as the percentage of effect (%) in the lowest concentration at which the event is still present on the evaluated population [20].

2.4. *Ames Test*

The method was applied according to Ames [45] using *Salmonella typhimurium* (*S. typhimurium*) TA98 and TA100 strains to evaluate possible mutagenicity. The cultures were grown in Oxoid nutrient broth No. 2. The samples of water were diluted in four different concentrations 25%, 50%, 75%, and 100% (*v/v*). The mutagenic effect was evaluated from the number of revertant colonies per plate. The plates were prepared in triplicate for every test sample, and the result presented was the mean of triplicate observation. The mutagenic activity was detected after 120 h of exposure at 37 (±2) °C. The revertant colonies' readings were counted using an automatic colony counter (Industrial Scientific). For accuracy of the results, mutagenic index (MI) values greater than or equal to two (≥2) were considered mutagenic [46].

2.5. Statistic Analysis

To establish whether there is a relationship between evaluated physicochemical parameters and heavy metals with results of bioassays, a one-way analysis of variance (ANOVA) was performed with a significance level of $p < 0.05$.

2.6. Microbiological Analysis

The determination of total coliforms and *E. coli* as indicators of bacterial fecal contamination was performed according to the ISO 9308-1 standard method [47]. Cellulose acetate membranes of 0.45 μm × 47 mm (Sartorius) were used for the filtration. Dark blue/purple colonies on Chromocult agar (Merck) were presumed to be *E. coli*. The detection and quantification of somatic coliphages as indicators of viral fecal contamination was performed according to the ISO 10705-2 standard procedure and modified Scholten's agar (MSA) (OXOID) was used for the detection of coliphages [48].

The results obtained in the samples from Village Gato, Village Tigui, and the water catchment of the Boque River were compared with normative 1594/1984 [49], which regulates waters that can be treated by conventional systems for human consumption. While for houses and deep-well underground sites, they were evaluated in compliance with normative 2115/2007 [38], which regulates water for human consumption. Normative 2115/2007 only beholds the microbiological quality concerning total coliforms and *E. coli*. Although coliphage concentrations are not regulated within Colombian normatives, their detection is relevant since they confirm contamination of fecal origin and the possible presence of pathogenic viruses, both in drinking water and in water for human consumption.

3. Results

3.1. Physicochemical Parameters

Parameters such as COD, total solids, and pH did not exceed the limits of Colombian normative 0631/2015 [37], in the first three sample stations. However, the level of cadmium was excessive in the Village Gato station in the first sampling with a concentration of 0.05 mg/L. Chromium did surpass the limit in the first sampling in Village Tigui. Likewise, mercury in Village Gato exceeded the limit in the third sampling (0.0029 mg/L). Moreover, in the Village Tigui station, the measurements in the second sampling exceeded 0.0025 mg/L and those in the water catchment of the Boque River in the first sampling were also excessive with 0.0022 mg/L. Furthermore, the permitted concentration of cyanides in the first and second sampling with a concentration of 1.02 and 1.32 mg/L, respectively, was excessive at the Village Tigui sampling station. While in the water catchment of the Boque River station, the prescribed cyanide level was exceeded in the second sampling with a concentration of 1.57 mg/L (Appendix A—Table A1).

The values of heavy metals analyzed in the last two sampling stations were compared with normative 2115/2007 [38] based on waters for human consumption. On one hand, the level of cadmium exceeded the limits established by the regulations for the house station in the second and third sampling with concentrations of 0.03 and 0.01 mg/L, respectively. While the cyanide concentration was exceeded only in the second sampling (1.11 mg/L). On the other hand, the established values of cadmium exceeded only in the second sampling in deep-well underground, presenting a concentration of 0.01 mg/L. Mercury was detected in each of the samples for both the house and deep-well underground stations, but the concentrations did not exceed the limits established in normative 2115 of 2007 [38]. The other metals evaluated (Zn, Ni, and Cr) were not detected in any of the samples analyzed (Appendix A—Table A2).

3.2. Bioassays

Table 1 shows the different percentages of growth inhibition of *L. sativa* in three samples for five evaluated stations. As the results show, there was root inhibition in some sampling stations and overgrowth in others. In Village Gato, in the third sampling, there was excess growth at a concentration of 25%. Likewise, in Village Tigui, the greatest inhibitions registered were observed at a concentration of 25%. In general, among all the sampled stations where the greatest inhibition was observed the highest was in Village Tigui, followed by house, deep-well underground, Village Gato, and the smallest recorded in the water catchment of the Boque River.

Table 1. Bioassays results with *Lactuca sativa*.

Sampling Station (n = 15)	**1S % (*v*/*v*) Effect**	**2S % (*v*/*v*) Effect**	**3S % (*v*/*v*) Effect**
Village Gato	33% Inhibition to 75%	27% Inhibition to 100%	133% Growth to 25%
Village Tigui	44% Inhibition to 25%	24% Inhibition to 25%	4% Inhibition to 50%
Water catchment of the Boque River	0% Inhibition to 100%	27% Inhibition to 100%	110% Growth to 100%
House	34% Inhibition to 25%	19% Inhibition to 75%	6% Inhibition to 50%
Deep-well underground	35% Inhibition to 50%	36% Inhibition to 75%	6% Inhibition to 75%

S: sampling, the numbers 1S, 2S, and 3S correspond to the months of July, September, and December in which the sample was taken; n: is the number of samples.

In Table 2, it is observed that the *D. magna* indicator has different mortality percentages since the same four concentrations of the sample are evaluated as in the *L. sativa* bioassay. The highest was 23% mortality at a concentration of 50% at the water catchment of the Boque River station and 17% mortality at 75% at the Village Tigui sampling station, followed by the house and deep-well underground stations. Finally, the lowest concentration of mortality was obtained in the Village Gato station with 33% mortality at 100%. For *H. attenuata*, it was not possible to determine the EC_{50} and LC_{50} values because there were no morphological changes, indicating lethality or sublethality, in the three samplings carried out, reporting 0% sublethality at 100% (*v/v*) and 0% lethality at 100% (*v/v*).

Table 2. Results of bioassay with *Daphnia magna*.

Sampling Station (n = 15)	**3S % (*v*/*v*) Effect**
Village Tigui	17% mortality to 75%
Village Gato	33% mortality to 100%
Water catchment of the Boque River	23% mortality to 50%
House	47% mortality to 100%
Deep-well underground	43% mortality to 100%

S: sampling, the number 3S corresponds to the month of December in which the sample was taken; n: is the number of samples.

3.3. Ames Test

The results obtained with the Ames test using *S. typhimurium* TA98 and TA100 strains are presented in Table 3, where the mutagenic index (MI) is shown for each condition used in the assay. According to Table 3, for the second sampling of the house station, mutagenic values were observed for both strain TA98 and TA100 in each of the concentrations evaluated. While for Village Gato, with strain TA100, a value of 2.4 was observed in the 100% concentration (Table 3). For the other sampling sites, there was no mutagenic index.

Table 3. Mutagenic index, for each concentration analyzed in the five sampling stations with strain *Salmonella typhimurium* TA98 and *S. typhimurium* TA100.

Sampling Stations (n = 15)	Concentration *v/v* (%)	TA98 (MI)			TA100 (MI)		
		S1	S2	S3	S1	S2	S3
Village Gato	25	0.77	1.07	0.00	0.97	0.46	1.00
	50	0.86	0.87	0.20	1.09	0.68	1.20
	75	1.06	1.49	0.60	1.27	0.77	1.70
	100	1.12	1.93	1.00	1.51	1.14	2.40
Village Tigui	25	0.97	0.70	0.00	1.37	0.20	0.60
	50	0.95	0.74	0.00	1.39	0.34	0.90
	75	1.12	1.14	0.10	1.48	0.33	1.00
	100	1.21	1.82	0.40	1.81	0.56	1.20
Water catchment of the Boque River	25	1.17	0.39	0.30	0.94	0.27	0.60
	50	1.39	0.63	0.50	1.00	0.24	0.70
	75	1.41	0.47	0.60	1.21	0.35	1.10
	100	1.50	0.59	0.90	1.71	0.44	1.30
House	25	0.68	41.78	0.40	1.01	11.05	0.60
	50	1.00	48.49	0.60	1.17	12.39	1.00
	75	1.06	56.62	0.80	1.36	13.75	2.00
	100	1.21	58.31	1.10	1.70	15.09	2.50
Deep-well underground	25	88.0	0.39	0.20	1.02	0.37	0.80
	50	0.88	0.50	0.50	1.16	1.21	1.20
	75	1.00	0.50	0.80	1.23	0.42	1.70
	100	0.55	0.64	0.90	1.28	0.64	2.00

S: sampling, the numbers 1S, 2S, and 3S correspond to the month of July, September, and December in which the sample was taken; n: is the number of samples; MI: mutagenic index.

3.4. Statistical Analysis

The statistical analysis performed to determine a possible relationship between the results of the physicochemical parameters against the toxicity indicators showed that there is a relationship between the inhibition of *L. sativa* concerning to mercury with a significance of $p < 0.05$.

However, due to the low number of samples analyzed for *H. attenuata* (three samples) versus the number of samples for *D. magna* (15 samples), it was not possible to establish whether there was a correlation with the concentration of metals or with the results of *L. sativa*.

3.5. Microbiological Analysis

Table 4 shows the results of the concentration of fecal contamination indicators (total coliforms, *E. coli*, and somatic coliphages) for the different types of water of the five sampling stations. Table 4 shows that concentrations between 10^3 and 10^5 colony forming unit (CFU)/100 mL for total coliforms were obtained at the different stations. While in the case of *E. coli*, concentrations between 10^3 and 10^4 CFU/100 mL were obtained. Somatic coliphages were detected in samples taken at Village Gato and house stations. Colombia does not have regulations for the presence of this indicator, although this is necessary since the presence of somatic coliphages represents a risk to the health of the community.

While, for the water catchment of the Boque River, Village Tigui, and deep-well underground stations, the presence of phages in some samples was not detected ($<1.0 \times 10^3$). The results were compared with decree 1594/1984 [49], which establishes the concentration of total coliforms (2.0×10^4/100 mL) allowed in waters that will be treated by conventional systems: while normative 2115/2007 [38], establishes that the concentrations for total coliforms and *E. coli* for drinking water is 0 CFU/100 mL.

Table 4. Results of total coliforms, *Escherichia coli*, and somatic coliphages in waters of the Boque River and drinking waters.

Sampling Stations (n = 15)	Microbiological Indicators								
	Total Coliforms CFU/100 mL			*E. coli* CFU/100 mL			Somatic Coliphages PFU/100 mL		
	1S	2S	3S	1S	2S	3S	1S	2S	3S
Village Gato	7.0×10^3	1.1×10^5	4.1×10^5	1.0×10^3	4.0×10^3	8.0×10^4	1.4×10^3	1.0×10^2	4.5
Village Tigui	1.3×10^5	3.2×10^4	2.2×10^5	2.0×10^3	2.0×10^3	3.0×10^4	4.9×10^3	3.0×10^2	$<1.0 \times 10^3$
Water catchment of the Boque River	2.4×10^4	1.7×10^4	4.0×10^5	1.0×10^3	1.0×10^3	1.0×10^4	$<1.0 \times 10^2$	$<1.0 \times 10^3$	1.9
House	4.0×10^4	1.4×10^5	2.8×10^5	1.0×10^3	3.0×10^4	3.0×10^4	1.0×10^2	2.0×10^4	1.0×10^2
Deep-well underground	3.2×10^4	6.0×10^4	3.8×10^5	1.0×10^3	4.0×10^4	4.0×10^4	2.0×10^2	$<1.0 \times 10^3$	1.0

CFU/100 mL: colony forming units in 100 mL of analyzed water; PFU/100 mL: plaque forming units in 100 mL of analyzed water; n: number of samples analyzed. S: sampling, the numbers 1S, 2S, and 3S correspond to the month of July, September, and December in which the sample was taken; <: less than the limit of quantification; n: is the number of samples.

4. Discussion

4.1. Bioassays

4.1.1. *Hydra attenuata* and *Daphnia magna*

By applying the *H. attenuata* toxicity test, it was not possible to determine lethality or sub-lethality since there were no morphological changes in the three samples taken. An important factor that could influence why *H. attenuata* was not sensitive to contaminants present in this water, is that the toxicity of metals is modified by abiotic factors such as hardness, pH, and water temperature [50]. For example, if water hardness is high, the formation of metal complexes tends to increase, which in turn lowers the effect of toxic divalent metals [50,51].

H. attenuata has a higher sensitivity to toxic substances at acidic pH, compared to that at alkaline or neutral pH [52]. The pH value of the water sample from the Boque River is about 7 (Appendix A—Tables A1 and A2), which could influence *H. attenuata* not presenting sensitivity when heavy metals, cyanides, or other toxic substances are in the water.

Due to the results obtained with *H. attenuata* in the first two samples, *D. magna* was used in the last sampling, to find an animal indicator that presented a greater sensitivity to these types of contaminants. Table 2 shows different mortality percentages that were found, demonstrating the sensitivity of this organism to the contaminants present in the water of the Boque River that is consumed by the Monterrey population. Studies conducted by Forget et al. [53] with *D. magna*, show percentages of toxicity up to 70% against heavy metals. Castro-Català et al. [54] evaluated the toxicity of sediments and water in rivers with the presence of pesticides and heavy metals using *D. magna* as an animal indicator, showing that it can be sensitive to these types of samples, due to its high metabolic rate [55]. Likewise, Lattuada et al. [56], in southern Brazil, used *D. magna* as an indicator of toxicity in waters affected by coal mining, in which heavy metals such as Fe, Mn, Zn, Ni, Cd, and Pb were found. The results showed sensitivity by this indicator in this type of water and suggest the evaluation of toxicity in waters from gold mining. Moreover, studies conducted in China by Wu et al. [57] demonstrated that the most frequently encountered heavy metals in a region affected by gold mining were mercury and cadmium, as observed in the results found in the drinking water of the population of Monterrey (Appendix A—Tables A1 and A2).

4.1.2. *Lactuca Sativa*

The differences observed between chemical parameters and toxicity may be related to the fact that the samples were not collected simultaneously and that it is not the same water because along the river route and on the different sampling days, diverse factors can alter its quality. Likewise, dilution effects due to rain, sedimentation, the introduction of new pollutants, among others, can have an influence.

Additionally, the water entering the treatment plant can be more contaminated, taking into account that it travels through tanks that are not in operation or comes into contact with sludge that might have a higher concentration of contaminants, which may return to the column of water.

The increase in germination, compared to the positive control (overgrowth), is related to the presence of organic matter because they are essential nutrients for *L. sativa* seed germination, and if they are available in high concentrations, they will stimulate growth. On the other hand, mercury and cyanide at the Village Tigui station (Appendix A—Tables A1 and A2) had higher concentrations. The inhibition rates of 24% and 44% for samples 1 and 2 at the concentration of 25%, affect the growth of the seed as the higher concentration of pollutants results in greater inhibition. Castillo et al. [24] found inhibition in the germination of seeds in waters contaminated with mercury and argue that it can occur due to the harmful effects caused by mercury at the cellular level in the seed. These results coincide with other studies where *L. sativa* has been proposed as a useful tool to evaluate and compare the toxicity of industrial effluents that present heavy metal contamination [58].

Likewise, the level of cadmium (Appendix A—Tables A1 and A2), also exceeded the minimum values established by the regulations for water for human consumption; studies have been reported where the exposure of *L. sativa* to this metal causes toxic and harmful effects that decrease its growth as the concentration of Cd, and thereby its adsorption, increases [59–61]. Just as the presence of heavy metals and cyanide has toxic implications in the plant and animal model, in the same way, it will affect the health of the human being [62,63].

Cd is one of the most toxic elements to which man is exposed since the accumulation of this metal in the body is gradual and increases with age due to its long half-life, greater than 20 years [64]. This is why eating food or drinking water with very high levels of cadmium causes severe stomach irritation, which causes vomiting, diarrhea, and sometimes death [64]. Moreover, cyanide exceeded the allowed limits (0.05 mg/L) in one of the samples analyzed in one of the houses (1.11 mg/L). The guide values of the World Health Organization [65] establish that the concentration of cyanide toxic to humans is 0.07 mg/L. Exposure to this concentration or higher may cause inhibition of cell growth, thereby affecting the breathing process and the metabolism of nitrogen and phosphate. It also inhibits the activity of some metalloproteins, joining cofactors such as the heme group of hemoglobin [66].

Finally, cyanide has acute effects on human health such as irritation of the eyes, nose, and throat. High exposure causes intoxication with headache, weakness, nausea, strong heartbeat, coma, and even death. As for chronic effects, it causes nosebleeds and nose lesions and can cause enlargement of the thyroid gland, which can interfere with its regular function [67].

4.2. Ames Test

In some cases, there was a decrease in reverts with increasing doses, which may be due to the presence of toxic substances that prevent the growth of bacteria [68]. However, in most of the sampling stations, a direct relationship was observed between the number of revertants and the increase in the concentration of heavy metals. This demonstrates the high probability of the presence of substances such as heavy metals and organic compounds in the Boque River that cause base-pair mutations and changes in the DNA reading frame of bacteria.

When observing the reversion of the strains, it was evidenced that they exceed 2–40 times the value of the negative control for the TA98 strain in the second sample in one of the houses, and from 2.0 to 2.5 for the TA100 strain in the third sample for the house and underground well. According to Orozco and Zuleta [69], some samples can exceed 100 times the negative control and these results are related to the quality of the water.

Likewise, Meléndez et al. [70] investigated the mutagenic activity of drinking water before and after chlorination at the Villa Hermosa plant, Medellín, Colombia, finding that contamination and chlorination influence mutagenicity. They used the Ames test with strains TA100 and TA98. Sierra et al. [71] evaluated the mutagenic activity of the Cauca River water with the same strains with and without the enzyme activator S9, finding that the highest rate of mutagenicity was observed with strain TA98 without enzyme activator. However, the TA100 strain is characterized by presenting the *hisG46* mutation and has specific markers that give it greater sensitivity to the test; within these are the *uvr* mutation, the *uvrB* mutation, and the plasmid pKM101 [26,68].

Mesquidaz et al. [72] reported alarming figures in the mercury concentrations used in the gold extraction process in a mine in northern Colombia, which ranges from 50 to 100 tons in 2007. Furthermore, it is reported that, thanks to this pollutant present in water, the health of the population has been affected, since Hg was found in human hair at a concentration of 12.8 μg/g, a figure that is well above international standards. It has been shown that inhabitants of different municipalities in southern Bolívar where gold mining takes place have high levels of Hg contamination, and this situation requires special attention to reduce environmental and human health problems [4]. Mercury contamination has been linked to health problems, as direct absorption of mercury vapor released by incinerators in gold mining, or ingestion of mercury-containing wastes, causes hydrargyrism and poisoning. Mercury (Hg) is one of the heavy metals of greatest concern to populations that consume fish. This pollutant can be released from many sources and has various toxic effects in humans [73].

Some of the health problems caused are excessive salivation, shortness of breath and fatigue, bronchitis, tremors and irritability, personality changes (due to brain damage), a sensation of floating teeth and pain in them, kidney and respiratory damage that can lead to death from problems in the lungs and other organs of the body [74–76]. While breathing polluted air, elemental mercury can reach the brain, affecting nerve cells and the olfactory system. The main organs in which mercury accumulates are the brain and the kidney [77,78].

4.3. Statistical Analysis

The statistical analysis showed a relationship between the inhibition of *L. sativa* concerning mercury with a significance of $p < 0.05$; this inhibition in the germination of *L. sativa* with this metal was also reported in Chile, where the exposure of the seeds to Hg inhibited their growth [19]. The toxicity caused multiple harmful effects in the seed at the cellular level such as a change in permeability in the cell membrane and the affinity to react with phosphate groups and the sulfhydryl group (SH). When mercury interacts with the SH groups to form the S–Hg–S bonds, it disrupts the stability of the group can affect seed germination and seedling growth whose tissues are rich in SH groups [79].

On the other hand, it was not possible to establish a correlation between the vegetal and the animal model due to the number of samples collected. When comparing the results of the bioassays associating for *D. magna* and *L. sativa*, it was observed that the variability due to the sampling was not simultaneous for every sample and it could be possibly affected by a new spill in the river. Additionally, these are different organisms with different sensitivity to the contaminants present in the Boque River water and there exist other factors that can influence this response. For example, some bacteria can naturally modify mercury (Hg^{2+}) by ion methylation forming CH_3–Hg^+, which is more toxic and is incorporated into trophic chains, affecting the animal model more than the vegetable model [80].

4.4. Total Coliforms, Escherichia Coli, and Somatic Coliphages

The microbiological results confirm the high fecal contamination in all the sampling stations (Table 4). Total coliform concentrations exceeded the limits for Colombian regulations [49]. In the case of drinking water for human consumption (deep-well underground and house), total coliforms and *E. coli* were well above levels required by regulation for drinking water [38]. Likewise, in the case of drinking water for human consumption by the treatment system (Village Gato, Tigui, and water catchment of the Boque River), the concentrations allowed for total coliforms were exceeded [49].

Campos-Pinilla et al. [81] and Sánchez-Alfonso et al. [82] in studies carried out in the Bogotá River found a total coliform concentration between 10^3 and 10^6 CFU/100 mL and for *E. coli* between 10^3 and 10^5/100 mL. This coincides with the values found in this study, which range between 10^3 and 10^5 CFU/100 mL of total coliforms and for *E. coli* between 10^3 and 10^4 CFU/100 mL (Table 4). Likewise, studies conducted by Lucena et al. [83] and Sánchez-Alfonso et al. [82] in rivers show average concentrations of somatic coliphages between 10^2 and 10^4 plaque forming unit (PFU)/100 mL, similar to those found in this study with ranges from 1 and 10^3 PFU/100 mL. The concentration of microorganisms in river water varies depending on climatic factors, geographical area, and the amount of organic matter present in water bodies [84,85]. The mine exploitation site is a settlement space for the population that works in this activity legally or illegally, which generates a high level of household waste in the river causing contamination by the discharge of fecal matter and organic matter, which explains the concentration of indicators of fecal contamination. It is related to the absence of treatment systems and improper installation of septic tanks.

The detected concentrations of total coliforms and *E. coli* in all the drinking water samples and the detection of somatic coliphages in some samples of water used for human consumption confirm the fecal contamination and the possible presence of pathogenic viruses in the drinking water (Table 4). These concentrations of indicators are similar to those detected in river samples as reported by Lucena et al. [83], Campos-Pinilla et al. [81], and Sánchez-Alfonso et al. [82], which could increase the risk for residents.

5. Conclusions

The results obtained with the three toxicity indicators reveal that *H. attenuata* does not present sensitivity to toxic substances present in this type of water, so its use for this purpose is not recommended. On the other hand, *D. magna* showed sensitivity even in diluted samples as well as *L. sativa*, which showed growth inhibition and excessive growth in different concentrations of the analyzed water, inclusive of waters with pollutant concentrations below the detection level. The Ames test shows an increase in the revertants indicating the possibility of mutations in the population that consumes this type of water, which is correlated with the results of the mutagenicity test that showed a mutagenic effect in the five stations evaluated with both strains used in the study. The highest mutagenic index was found in the water sample taken from the house sampling station. The concentration of bacteria in the water exceeded the limits allowed by Colombian regulations, creating a health risk, also with an alert call to the presence of possible pathogenic viruses, and the risk that they imply for the inhabitants of Monterrey due to somatic coliphage levels determined.

This research recognizes the potential use of bioassays to evaluate the toxic effects generated by chemical wastes produced by gold mining and discharged into surface waters. The use of animal and plant models is recommended to evaluate said effects on the environment and public health and infer the damages that until now have not been sufficiently evaluated having as correlation factors physicochemical and microbiological parameters.

Finally, this research generated data that contribute to the knowledge of the effects caused to the environmental and public health by illegal and legal mining carried out with bad practices in emerging countries with inefficient controls of this type of activity. These assays used can help sanitation organizations in different countries to take preventive actions on this issue

Author Contributions: All authors contributed to all features of the paper. A.M., J.A., J.L., L.S., C.V., A.O.-A., and M.D. were involved in the sampling and analysis of bacteria, coliphages, bioassays, and physicochemical determinations. C.C. and C.C.Z. conceived the idea for the research and contributed to the development of the project by obtaining economic resources. C.C., N.d.P., and C.C.Z. provided consultation and interpretation of the results and contributed to writing the manuscript. C.V., A.M., and J.L. edited the manuscript. All authors have read and agreed to the published version of the manuscript.

Funding: This research and article publication was funded by Pontificia Universidad Javeriana, Bogotá, Colombia. Grant number 0005746 and 0008469.

Acknowledgments: The authors would like to acknowledgment the residents of Monterrey Sur de Bolívar, Colombia, and Programa de Desarrollo y Paz del Magdalena Medio by the logistical support in sampling.

Conflicts of Interest: The authors declare no conflict of interest.

Appendix A

Table A1. Results of physicochemical parameter analysis compared with the normative 0631/2015.

Physicochemical Parameters (n = 9)	Village Gato			Village Tigui			Water Catchment of the Boque River			Limit of the Regulations
Number of Sampling	**S1**	**S2**	**S3**	**S1**	**S2**	**S3**	**S1**	**S2**	**S3**	**Normative 0631/2015 [37]**
pH	7.36	7.49	7.33	7.55	6.05	6.57	7.61	7.56	7.63	6.0–9.0
COD (mg/L)	32.67	<0.001	5.52	22.46	<0.001	24.17	29.63	40.72	4.82	150
Total solids (g/10 mL)	0.0024	0.00165	0.0	0.0007	0.0058	0.0029	0.0001	0.00316	0.0005	50
Cyanide (mg/L)	<0.025	<0.025	0.025	1.02	1.32	<0.025	<0.025	1.57	<0.025	1.0
Cadmium (Cd) (mg/L)	0.05	0.03	0.02	$<1.0 \times 10^{-2}$	0.02	0.02	$<1.0 \times 10^{-2}$	0.02	0.01	0.05
Chrome (Cr) (mg/L)	$<1.0 \times 10^{-6}$	$<1.0 \times 10^{-6}$	$<1.0 \times 10^{-6}$	0.06	$<1.0 \times 10^{-6}$	0.04	$<1.0 \times 10^{-6}$	$<1.0 \times 10^{-6}$	$<1.0 \times 10^{-6}$	0.5
Mercury (Hg) (mg/L)	0.0008	0.0029	0.0003	0.001	0.0025	0.0008	0.0022	0.002	0.0008	0.002
Nickel (Ni) (mg/L)	$<1.0 \times 10^{-3}$	$<1.0 \times 10^{-3}$	$<1.0 \times 10^{-3}$	$<1.0 \times 10^{-3}$	$<1.0 \times 10^{-3}$	$<1.0 \times 10^{-3}$	$<1.0 \times 10^{-3}$	$<1.0 \times 10^{-3}$	$<1.0 \times 10^{-3}$	0.5
Zinc (Zn) (mg/L)	$<1.0 \times 10^{-3}$	$<1.0 \times 10^{-3}$	$<1.0 \times 10^{-3}$	$<1.0 \times 10^{-3}$	$<1.0 \times 10^{-3}$	$<1.0 \times 10^{-3}$	$<1.0 \times 10^{-3}$	$<1.0 \times 10^{-3}$	$<1.0 \times 10^{-3}$	3.0

mg/L: milligram per liter; S: sampling, the numbers 1S, 2S, and 3S correspond to the month of July, September, and December in which the sample was taken; <: less than the limit of quantification; n: is the number of samples; COD: chemical oxygen demand.

Table A2. Results of physicochemical parameters analysis compared with the normative 2115 of 2007.

Physicochemical Parameters (n = 6)	House			Deep-Well Underground			Limit of the Regulations
Number of sampling	**S1**	**S2**	**S3**	**S1**	**S2**	**S3**	**Normative 2115/2007 [38]**
pH	7.6	7.53	7.44	6.95	6.79	6.61	6.5–9.0
Cyanide (mg/L)	<0.025	1.11	<0.025	<0.025	<0.025	<0.025	0.05
Cadmium (Cd) (mg/L)	$<1.0 \times 10^{-2}$	0.03	0.03	$<1.0 \times 10^{-2}$	0.01	$<1.0 \times 10^{-2}$	0.003
Chrome (Cr) (mg/L)	0.02	$<1.0 \times 10^{-6}$	$<1 \times 10^{-6}$	$<1.0 \times 10^{-6}$	$<1.0 \times 10^{-6}$	$<1.0 \times 10^{-6}$	0.05
Mercury (Hg) (mg/L)	0.0004	0.0005	0.0003	0.0007	0.0003	0.0007	0.001
Nickel (Ni) (mg/L)	$<1.0 \times 10^{-3}$	$<1.0 \times 10^{-3}$	$<1.0 \times 10^{-3}$	$<1.0 \times 10^{-3}$	$<1.0 \times 10^{-3}$	$<1.0 \times 10^{-3}$	0.2
Zinc (Zn) (mg/L)	$<1.0 \times 10^{-3}$	$<1.0 \times 10^{-3}$	$<1.0 \times 10^{-3}$	$<1.0 \times 10^{-3}$	$<1.0 \times 10^{-3}$	$<1.0 \times 10^{-3}$	3.0

mg/L: milligram per liter; S: sampling, the numbers 1S, 2S, and 3S correspond to the month of July, September, and December in which the sample was taken; <: less than the limit of quantification; n: is the number of samples.

References

1. UNIDO Green Industry Initiative for Sustainable Industrial Development 2011. Available online: https://www.greengrowthknowledge.org/sites/default/files/downloads/resource/Green_Industry_Initiative_for_Sustainable_Development_UNIDO.pdf (accessed on 12 February 2020).
2. Ogola, J.S.; Mitullah, W.V.; Omulo, M.A. Impact of gold mining on the environment and human health: A case study in the Migori Gold Belt, Kenya. *Environ. Geochem. Health* **2002**, *24*, 141–157. [CrossRef]
3. Gafur, N.A.; Sakakibara, M.; Sano, S.; Sera, K. A case study of heavy metal pollution in water of Bone River by artisanal small-scale gold mine activities in eastern part of Gorontalo, Indonesia. *Water* **2018**, *10*, 1507. [CrossRef]
4. Carranza-Lopez, L.; Caballero-Gallardo, K.; Cervantes-Ceballos, L.; Turizo-Tapia, A.; Olivero-Verbel, J. Multicompartment mercury contamination in major gold mining districts at the department of Bolivar, Colombia. *Arch. Environ. Contam. Toxicol.* **2019**, *76*, 640–649. [CrossRef] [PubMed]
5. Pinedo-Hernández, J.; Marrugo-Negrete, J.; Díez, S. Speciation and bioavailability of mercury in sediments impacted by gold mining in Colombia. *Chemosphere* **2015**, *119*, 1289–1295. [CrossRef] [PubMed]
6. Von Behren, J.; Liu, R.; Sellen, J.; Duffy, C.N.; Gajek, R.; Choe, K.-Y.; DeGuzman, J.; Janes, M.K.; Hild, J.; Reynolds, P. Heavy metals in California women living in a gold mining-impacted community. *Int. J. Environ. Res. Public Health* **2019**, *16*, 2252. [CrossRef]
7. Ouboter, P.E.; Landburg, G.; Satnarain, G.U.; Starke, S.Y.; Nanden, I.; Simon-Friedt, B.; Hawkins, W.B.; Taylor, R.; Lichtveld, M.Y.; Harville, E.; et al. Mercury levels in women and children from interior villages in Suriname, South America. *Int. J. Environ. Res. Public Health* **2018**, *15*, 1007. [CrossRef] [PubMed]
8. Producción de oro en Colombia Creció 7% en el Primer Trimestre de 2020. Available online: https://www.elespectador.com/economia/produccion-de-oro-en-colombia-crecio-7-en-el-primer-trimestre-de-2020-articulo-920587/ (accessed on 22 May 2020).
9. Robles Mengoa, M.E.; Urán, A. Colombia: Legal loopholes behind illegal gold trade. In *Global Gold Production Touching Ground*, 1st ed.; Springer International Publishing: New York, NY, USA, 2020; pp. 151–161.
10. Güiza, L. Small scale mining in Colombia: Not such a small activity. *DYNA* **2013**, *80*, 109–117.
11. DANE Proyecciones de Población Departamentales Por Área 2015–2020. Available online: https://www.dane.gov.co/files/investigaciones/poblacion/proyepobla06_20/7Proyecciones_poblacion.pdf (accessed on 22 February 2020).
12. Ministerio de Minas y Energía Censo Minero Departamental 2010–2011. Available online: https://www.minenergia.gov.co/documents/10180/698204/CensoMinero.pdf/093cec57-05e8-416b-8e0c-5e4f7c1d6820 (accessed on 8 February 2020).
13. De la Hoz, V.J. Economía y conflicto en el Cono Sur del Departamento de Bolívar. *Banco de la República* **2009**, *110*, 1–109.
14. Alcaldía de Simití-Bolívar Plan de Desarrollo ¡Simití Pata Todos! 2016–2019. Available online: https://simitibolivar.micolombiadigital.gov.co/sites/simitibolivar/content/files/000132/6580_plan-de-desarrollo-municipal-20162019.pdf (accessed on 22 July 2020).
15. Grajales, A.K. *El Río Grande la Magdalena Como Eje Estructurante del Desarrollo territorial. Estudio de Caso: Magdalena Medio 1950–2010*; Empresas Publicas de Medellin (EPM): Medellin, Colombia, 2012; pp. 1–65.
16. Donato, D.B.; Nichols, O.; Possingham, H.; Moore, M.; Ricci, P.F.; Noller, B.N. A critical review of the effects of gold cyanide-bearing tailings solutions on wildlife. *Environ. Int.* **2007**, *33*, 974–984. [CrossRef]
17. Hilson, G. Abatement of mercury pollution in the small-scale gold mining industry: Restructuring the policy and research agendas. *Sci. Total Environ.* **2006**, *362*, 1–14. [CrossRef]
18. Hilson, G.; Murck, B. Progress toward pollution prevention and waste minimization in the North American gold mining industry. *J. Clean Prod.* **2001**, *9*, 405–415. [CrossRef]
19. Cordy, P.; Veiga, M.M.; Salih, I.; Al-Saadi, S.; Console, S.; Garcia, O.; Mesa, L.A.; Velásquez-López, P.C.; Roeser, M. Mercury contamination from artisanal gold mining in Antioquia, Colombia: The world's highest per capita mercury pollution. *Sci. Total Environ.* **2011**, *410*, 154–160. [CrossRef] [PubMed]
20. Bohórquez-Echeverry, P.; Duarte-Castañeda, M.; León-López, N.; Caicedo-Carrascal, F.; Vásquez-Vásquez, M.; Campos-Pinilla, C. Selection of a bioassay battery to assess toxicity in the affluents and effluents of three water-treatment plants. *Univ. Sci.* **2012**, *17*, 152–166. [CrossRef]

21. Keddy, C.J.; Greene, J.C.; Bonnell, M.A. Review of whole-organism bioassays: Soil, freshwater sediment, and freshwater assessment in Canada. *Ecotoxicol. Environ. Saf.* **1995**, *30*, 221–251. [CrossRef]
22. Miller, W.E.; Peterson, S.A.; Greene, J.C.; Callahan, C.A. Comparative toxicology of laboratory organisms for assessing Hazardous Waste Sites. *J. Environ. Qual.* **1985**, *14*, 569–574. [CrossRef]
23. Férard, J.F.; Burga Pérez, K.F.; Blaise, C.; Péry, A.; Sutthivaiyakit, P.; Gagné, F. Microscale ecotoxicity testing of Moselle river watershed (Lorraine Province, France) sediments. *J. Xenobiot.* **2015**, *5*, 1–7. [CrossRef]
24. Castillo, G.C.; Vila, I.C.; Ella, N. Ecotoxicity assessment of metals and wastewater using multitrophic assays. *Environ. Toxicol.* **2000**, *15*, 370–375. [CrossRef]
25. MacGregor, J.T.; Casciano, D.; Müller, L. Strategies and testing methods for identifying mutagenic risks. *Mutat. Res.* **2000**, *455*, 3–20. [CrossRef]
26. Maron, D.M.; Ames, B.N. Revised methods for the *Salmonella* mutagenicity test. *Mutat. Res.* **1983**, *113*, 173–215. [CrossRef]
27. González Scancella, T. *Evaluación del Sistema de Abastecimiento de Agua Potable y Disposición de Excretas de la Población del Corregimiento de Monterrey, Municipio de Simití, Departamento de Bolívar, Proponiendo Soluciones Integrales al Mejoramiento de Los Sistemas y la Salud de la Comunidad*; Ecology College Degree, Pontificia Universidad Javeriana: Bogotá, Colombia, 2013.
28. Barrera, J.A.; Espinosa, A.J.; Álvarez, J.P. Contaminación en el Lago de Tota, Colombia: Toxicidad aguda en Daphnia magna (Cladocera: Daphniidae) e Hydra attenuata (Hydroida: Hydridae). *Rev. biol. Trop.* **2019**, *67*, 11–23. [CrossRef]
29. Espinosa-Ramírez, A.J. El Agua, un Reto Para la Salud Pública. La Calidad del Agua y las Oportunidades Para la Vigilancia en Salud Ambiental. Ph.D. Thesis, Universidad Nacional de Colombia, Bogotá, Colombia, 2018.
30. Zeiger, E. The test that changed the world: The Ames test and the regulation of chemicals. *Mutat. Res.* **2019**, *841*, 43–48. [CrossRef] [PubMed]
31. American Public Health Association 2310 B. Titration method. In *Standard Methods for the Examination of Water and Wastewater*; American Public Health Association: Washington, DC, USA, 2012; p. 1946.
32. American Public Health Association 5220 D. Closed reflux, colorimetric method. In *Standard Methods for the Examination of Water and Wastewater*; American Public Health Association: Washington, DC, USA, 2012; p. 1946.
33. American Public Health Association 2540 B. Total solids dried at 103–105 °C. In *Standard Methods for the Examination of Water and Wastewater*; American Public Health Association: Washington, DC, USA, 2012; p. 1946.
34. American Public Health Association 3111. Metals by flame atomic absorption spectrometry. In *Standard Methods for the Examination of Water and Wastewater*; American Public Health Association: Washington, DC, USA, 2012; p. 1946.
35. American Public Health Association 3112B. Metals by cold-vapor atomic absorption spectrometry. In *Standard Methods for the Examination of Water and Wastewater*; American Public Health Association: Washington, DC, USA, 2012; p. 1946.
36. American Public Health Association 4500-CN-E. Colorimetric method. In *Standard Methods for the Examination of Water and Wastewater*; American Public Health Association: Washington, DC, USA, 2012; p. 1946.
37. Ministerio de Ambiente y Desarrollo Sostenible Resolución 0631 de 2015, Parámetros y los Valores Límites Máximos Permisibles en los Vertimientos Puntuales a Cuerpos de Aguas Superficiales y a Los Sistemas de al Cantarillado Público y se Dictan Otras Disposiciones. Available online: https://www.minambiente.gov.co/images/normativa/app/resoluciones/d1-res_631_marz_2015.pdf (accessed on 21 March 2020).
38. Ministerio de Protección Social, Ministerio de Ambiente, Vivienda y Desarrollo Territorial Resolución 2115 de 2007, Características, Instrumentos Básicos y Frecuencias del Sistema de Control y Vigilancia Para la Calidad del Agua Para Consumo Humano. Available online: https://www.minambiente.gov.co/images/GestionIntegraldelRecursoHidrico/pdf/Legislaci%C3%B3n_del_agua/Resoluci%C3%B3n_2115.pdf (accessed on 21 March 2020).
39. Dutka, B. Short-term root elongation toxicity bioassay. In *Methods for Toxicological Analysis of Waters, Wastewaters and Sediments, Rivers Research Branch*; Canada Centre for Inland Waters, National Water Research Institute: Burlington, ON, Canada, 1989; pp. 120–122.
40. Trottier, S.; Biaise, C.; Kusui, T.; Johnson, E.M. Acute toxicity assessment of aqueous samples using a microplate-based *hydra attenuata* assay. *Environ. Toxicol. Water Qual.* **1997**, *12*, 265–271. [CrossRef]

41. Dutka, B. Daphnia magna 48 hours static bioassay method for acute toxicity in environmental samples. In *Methods for Toxicological Analysis of Waters, Wastewaters and Sediments*; Rivers Research Branch; Canada Centre for Inland Waters, National Water Research Institute: Burlington, ON, Canada, 1989; pp. 55–59.
42. EPA. *Method for Measuring the Acute Toxicity of Effluents to Freshwater and Marine Organism*, 3rd ed.; Environmental Monitoring Systems Laboratory, Office of Research and Development Environmental Protection Agency: Washington, DC, USA, 1985; p. 231.
43. Finney, D.J. *Probit Analysis*, 2nd ed.; Cambridge University Press: New York, NY, USA, 1952; p. 318.
44. Hamilton, M.A.; Russo, R.C.; Thurston, R.V. Trimmed spearman-karber method for estimating median lethal concentrations in toxicity bioassays. *Environ. Sci. Technol.* **1977**, *11*, 714–719. [CrossRef]
45. Williams, L.; Preston, J.E. *Interim Procedures for Conducting the Salmonella/Microsomal Mutagenicity Assay (Ames Test)*; Environmental Protection Agency: Las Vegas, NV, USA, 1983; p. 40.
46. Ames, B.N.; Kammen, H.O.; Yamasaki, E. Hair Dyes are mutagenic: Identification of a variety of mutagenic ingredients. *Proc. Natl. Acad. Sci. USA* **1975**, *72*, 2423–2427. [CrossRef] [PubMed]
47. International Organization for Standardization. *ISO 9308-1: Water Quality—Enumeration of Escherichia coli and Coliform Bacteria-Part 1: Membrane Filtration Method for Waters with Low Bacterial Background Flora*, 3rd ed.; International Organization for Standardization: Geneva, Switzerland, 2014.
48. International Organization for Standardization. *ISO 10705-2: Water Quality. Detection and Enumeration of Bacteriophages-Part 2: Enumeration of Somatic Coliphages*, 1st ed.; International Organization for Standardization: Geneva, Switzerland, 2000.
49. de Ministerio, A. Decreto 1594 de 1984, Por el cual se Reglamenta Parcialmente el del Decreto Ley 2811 de 1974 en Cuanto a Usos del Agua y Residuos Líquidos. Available online: https://www.minambiente.gov.co/images/GestionIntegraldelRecursoHidrico/pdf/normativa/Decreto_1594_de_1984.pdf (accessed on 6 February 2020).
50. Holdway, D.A.; Lok, K.; Semaan, M. The acute and chronic toxicity of cadmium and zinc to two *hydra* species. *Environ. Toxicol.* **2001**, *16*, 557–565. [CrossRef]
51. Riethmuller, N.; Markich, S.J.; Van Dam, R.A.; Parry, D. Effects of water hardness and alkalinity on the toxicity of uranium to a tropical freshwater hydra (*hydra viridissima*). *Biomarkers* **2001**, *6*, 45–51. [CrossRef]
52. Sundaram, R.; Smith, B.W.; Clark, T.M. pH-dependent toxicity of serotonin selective reuptake inhibitors in taxonomically diverse freshwater invertebrate species. *Mar. Freshw. Res.* **2015**, *66*, 518–525. [CrossRef]
53. Forget, G.; Gagnon, P.; Sanchez, W.A.; Dutka, B.J. Overview of methods and results of the eight country International Development Research Centre (IDRC) WaterTox project. *Environ. Toxicol.* **2000**, *15*, 264–276. [CrossRef]
54. de Castro-Català, N.; Kuzmanovic, M.; Roig, N.; Sierra, J.; Ginebreda, A.; Barceló, D.; Pérez, S.; Petrovic, M.; Picó, Y.; Schuhmacher, M.; et al. Ecotoxicity of sediments in rivers: Invertebrate community, toxicity bioassays and the toxic unit approach as complementary assessment tools. *Sci. Total Environ.* **2016**, *540*, 297–306. [CrossRef]
55. Yuan, N.; Pei, Y.; Bao, A.; Wang, C. The Physiological and Biochemical Responses of *Daphnia magna* to Dewatered Drinking Water Treatment Residue. *Int. J. Environ. Res. Public Health* **2020**, *17*, 5863. [CrossRef]
56. Lattuada, R.M.; Menezes, C.T.B.; Pavei, P.T.; Peralba, M.C.R.; Dos Santos, J.H.Z. Determination of metals by total reflection X-ray fluorescence and evaluation of toxicity of a river impacted by coal mining in the south of Brazil. *J. Hazard. Mater.* **2009**, *163*, 531–537. [CrossRef] [PubMed]
57. Wu, Y.G.; Xu, Y.N.; Zhang, J.H.; Hu, S.H. Evaluation of ecological risk and primary empirical research on heavy metals in polluted soil over Xiaoqinling gold mining region, Shaanxi, China. *Trans. Nonferrous Met. Soc. China* **2010**, *20*, 688–694. [CrossRef]
58. Charles, J.; Sancey, B.; Morin-Crini, N.; Badot, P.M.; Degiorgi, F.; Trunfio, G.; Crini, G. Evaluation of the phytotoxicity of polycontaminated industrial effluents using the lettuce plant (*Lactuca sativa*) as a bioindicator. *Ecotoxicol Environ. Saf.* **2011**, *74*, 2057–2064. [CrossRef]
59. Cornu, J.Y.; Denaix, L.; Schneider, A.; Pellerin, S. Temporal variability of solution Cd^{2+} concentration in metal-contaminated soils as affected by soil temperature: Consequences on lettuce (*Lactuca sativa* L.) exposure. *Plant. Soil* **2008**, *307*, 51–65. [CrossRef]
60. Tang, X.; Pang, Y.; Ji, P.; Gao, P.; Nguyen, T.H.; Tong, Y. Cadmium uptake in above-ground parts of lettuce (*Lactuca sativa* L.). *Ecotoxicol. Environ. Saf.* **2016**, *125*, 102–106. [CrossRef] [PubMed]

61. Retamal-Salgado, J.; Hirzel, J.; Walter, I.; Matus, I. Bioabsorption and Bioaccumulation of Cadmium in the Straw and Grain of Maize (*Zea mays* L.) in Growing Soils Contaminated with Cadmium in Different Environment. *Int. J. Environ. Res. Public Health* **2017**, *14*, 1399. [CrossRef]
62. Duri, L.; Visconti, D.; Fiorentino, N.; Adamo, P.; Fagnano, M.; Caporale, A. Health risk assessment in agricultural soil potentially contaminated by Geogenic Thallium: Influence of plant species on metal mobility in soil-plant system. *Agronomy* **2020**, *10*, 890. [CrossRef]
63. Mercado-Garcia, D.; Beeckman, E.; Van Butsel, J.; Arroyo, N.D.; Peña, M.S.; Van Buggendhoudt, C.; Saeyer, N.D.; Forio, M.A.E.; Schamphelaere, K.A.C.D.; Wyseure, G.; et al. Assessing the freshwater quality of a large-scale miningwatershed: The need for integrated approaches. *Water* **2019**, *11*, 1797. [CrossRef]
64. Atsdr, E.P.A. Agency for toxic substances and disease registry case studies in environmental medicine (CSEM). *Cadmium* **2011**, *1*, 1–63.
65. Guidelines for Drinking-Water Quality. Available online: https://www.paho.org/en/documents/guidelines-drinking-water-quality-4o-ed-2011 (accessed on 19 March 2020).
66. Castillo Rodriguez, F. *Biotecnología Ambiental*; Editorial Tebar Flores: Madrid, Spain, 2005; p. 592.
67. New Jersey Department of Health and Senior Services Hazardous Substance Fact Sheet 2007. Available online: http://www.nj.gov/health/eoh/rtkweb/documents/fs/0027.pdf (accessed on 22 March 2020).
68. Ames, B.N.; Lee, F.D.; Durston, W.E. An improved bacterial test system for the detection and classification of Mutagens and Carcinogens. *Proc. Natl. Acad. Sci. USA* **1973**, *70*, 782–786. [CrossRef]
69. Orozco, L.Y.; Zuleta, M. Detección de mutacarcinógenos en aguas del Río Pantanillo y efecto genotóxico de esta agua en el DNA nuclear y mitocondrial de células eucarióticas. *Iatreia* **2000**, *13*, 94.
70. Meléndez, I.; Zuleta, M.; Marín, I.; Calle, J.; Salazar, D. Actividad mutagénica de aguas de consumo humano antes y después de clorar en la planta de Villa Hermosa, Medellín. *Iatreia* **2001**, *14*, 167–175.
71. Sierra, J.; Benitez, N.; Bravo, E.; Larmat, F.; Soto, A. Evaluation of the mutagenic activity of waters collected from the Cauca River in the city Cali, Colombia by using the *Salmonella*/Microsome Assay. *Revista de Ciencias* **2012**, *16*, 131.
72. Mesquidaz, E.D.; Negrete, J.M.; Hernández, J.P. Exposición a mercurio en trabajadores de una mina de oro en el norte de Colombia. *Salud Uninorte* **2013**, *29*, 534–541.
73. Manjarres-Suarez, A.; Olivero-Verbel, J. Hematological parameters and hair mercury levels in adolescents from the Colombian Caribbean. *Environ. Sci. Pollut. Res.* **2020**, *27*, 14216–14227. [CrossRef]
74. Zahir, F.; Rizwi, S.J.; Haq, S.K.; Khan, R.H. Low dose mercury toxicity and human health. *Environ. Toxicol Pharmacol.* **2005**, *20*, 351–360. [CrossRef]
75. Næss, S.; Kjellevold, M.; Dahl, L.; Nerhus, I.; Midtbø, L.K.; Bank, M.S.; Rasinger, J.D.; Markhus, M.W. Effects of seafood consumption on mercury exposure in Norwegian pregnant women: A randomized controlled trial. *Environ. Int.* **2020**, *141*, 105759. [CrossRef]
76. Pateda, S.; Sakakibara, M.; Sera, K. Lung function assessment as an early biomonitor of mercury-induced health disorders in artisanal and small-scale gold mining areas in Indonesia. *Int. J. Environ. Res. Public Health* **2018**, *15*, 2480. [CrossRef]
77. Park, J.-D.; Zheng, W. Human exposure and health effects of inorganic and elemental mercury. *J. Prev. Med. Public Health* **2012**, *34445*, 344–352. [CrossRef]
78. Gyamfi, O.; Sorenson, P.B.; Darko, G.; Ansah, E.; Bak, J.L. Human health risk assessment of exposure to indoor mercury vapour in a Ghanaian artisanal small-scale gold mining community. *Chemosphere* **2020**, *241*, 125014. [CrossRef]
79. Patra, M.; Bhowmik, N.; Bandopadhyay, B.; Sharma, A. Comparison of mercury, lead and arsenic with respect to genotoxic effects on plant systems and the development of genetic tolerance. *Environ. Exp. Bot.* **2004**, *52*, 199–223. [CrossRef]
80. Clarkson, T.W.; Magos, L. The toxicology of mercury and its chemical compounds. *Crit. Rev. Toxicol.* **2006**, *36*, 609–662. [CrossRef] [PubMed]
81. Campos-Pinilla, C.; Cárdenas-Guzmán, M.; Guerrero-Cañizares, A. Comportamiento de los indicadores de contaminación fecal en diferente tipo de aguas de la sabana de Bogotá (Colombia). *Univ. Sci.* **2008**, *13*, 103–108.
82. Sánchez-Alfonso, A.C.; Venegas, C.; Díez, H.; Méndez, J.; Blanch, A.R.; Jofre, J.; Campos, C. Microbial indicators and molecular markers used to differentiate the source of faecal pollution in the Bogotá River (Colombia). *Int. J. Hyg. Environ. Health* **2020**, *225*, 113450. [CrossRef] [PubMed]

83. Lucena, F.; Méndez, X.; Morón, A.; Calderón, E.; Campos, C.; Guerrero, A.; Cárdenas, M.; Gantzer, C.; Shwartzbrood, L.; Skraber, S.; et al. Occurrence and densities of bacteriophages proposed as indicators and bacterial indicators in river waters from Europe and South America. *J. Appl. Microbiol.* **2003**, *94*, 808–815. [CrossRef]
84. Augustyn, Ł.; Babula, A.M.; Joniec, J.; Stanek-Tarkowska, J.; Hajduk, E.; Kaniuczak, J. Microbiological indicators of the quality of river water, used for drinking water supply. *Pol. J. Environ. Stud.* **2016**, *25*, 511–519. [CrossRef]
85. Jofre, J.; Lucena, F.; Blanch, A.R.; Muniesa, M. Coliphages as model organisms in the characterization and management of water resources. *Water* **2016**, *8*, 199. [CrossRef]

Article

Biodegradation of Amoxicillin, Tetracyclines and Sulfonamides in Wastewater Sludge

Chu-Wen Yang, Chien Liu and Bea-Ven Chang *

Department of Microbiology, Soochow University, Taipei 11101, Taiwan; ycw6861@scu.edu.tw (C.-W.Y.); jennifer719123@yahoo.com.tw (C.L.)

* Correspondence: bvchang@scu.edu.tw; Tel.: +886-228-819-471 (ext. 6859)

Received: 22 May 2020; Accepted: 28 July 2020; Published: 30 July 2020

Abstract: The removal of antibiotics from the aquatic environment has received great interest. The aim of this study is to examine degradation of oxytetracycline (OTC), tetracycline (TC), chlortetracycline (CTC), amoxicillin (AMO), sulfamethazine (SMZ), sulfamethoxazole (SMX), sulfadimethoxine (SDM) in sludge. Four antibiotic-degrading bacterial strains, SF1 (*Pseudmonas* sp.), A12 (*Pseudmonas* sp.), strains B (*Bacillus* sp.), and SANA (*Clostridium* sp.), were isolated, identified and tested under aerobic and anaerobic conditions in this study. Batch experiments indicated that the addition of SF1 and A12 under aerobic conditions and the addition of B and SANA under anaerobic conditions increased the biodegradation of antibiotics in sludge. Moreover, the results of repeated addition experiments indicated that the efficiency of the biodegradation of antibiotics using the isolated bacterial strains could be maintained for three degradation cycles. Two groups of potential microbial communities associated with the aerobic and anaerobic degradation of SMX, AMO and CTC in sludge were revealed. Twenty-four reported antibiotics-degrading bacterial genera (*Achromobacter, Acidovorax, Acinetobacter, Alcaligenes, Bacillus, Burkholderia, Castellaniella, Comamonas, Corynebacterium, Cupriavidus, Dechloromonas, Geobacter, Gordonia, Klebsiella, Mycobacterium, Novosphingobium, Pandoraea, Pseudomonas, Rhodococcus, Sphingomonas, Thauera, Treponema, Vibrio* and *Xanthobacter*) were found in both the aerobic and anaerobic groups, suggesting that these 24 bacterial genera may be the major antibiotic-degrading bacteria in sludge.

Keywords: sludge; antibiotics; biodegradation

---

1. Introduction

Antibiotics are discharged from animals (and humans) and enter wastewater treatment plants (WWTPs) through the sewage system [1]. In general, antibiotics are not easy for wastewater treatment facilities to remove [2]. The release and persistence of antibiotics in the environments may lead to an increase in antibiotic-resistant bacteria [3]. Tetracycline antibiotics (TCs), such as tetracycline (TC), oxytetracycline (OTC), and chlortetracycline (CTC), are broad-spectrum antibacterial drugs that inhibit bacterial protein synthesis [4]. Sulfonamide antibiotics (SAs) act against most Gram-positive and many Gram-negative bacteria. SAs act as competitive inhibitors of *p*-aminobenzoic acid in the folic acid synthetic pathway to inhibit the growth of bacteria [5]. Three sulfonamides, sulfamethoxazole (SMX), sulfadimethoxine (SDM), and sulfamethazine (SMZ), are found in sludge in many WWTPs and rivers [6–8]. Amoxicillin (AMO) is a β-lactam family antibiotic drug. This class of antibiotics disrupts bacterial cell walls during bacterial growth [9]. AMO in the environment may lead to an increase in antibiotic-resistant bacteria [10].

Several methods for antibiotic removal from wastewater sludge have been proposed. These approaches include chemical oxidation methods such as UV- and solar-based procedures (UV/H_2O_2, solar/H_2O_2), ozonation, photocatalysis by Fe^{2+} or Fe^{3+}/H_2O_2 and TiO_2 photocatalysis [11].

Antibiotic degradation by fungal extracellular enzymes extracted from spent mushroom composts has also been reported [12]. Thermal hydrolysis/anaerobic digestion [13] and the use of aerobic granular sludge treated with manganese oxides [14,15] have also been proposed. Nevertheless, many antibiotics continue to be detected in wastewater sludge [16–18]. The use of microorganisms to eliminate antibiotics is a promising strategy [19–21]. The aerobic degradation of antibiotics has been observed in pure bacterial cultures and microbial consortiums. *Rhodococcus rhodochrous* and *Aspergillus niger* degrade pharmaceuticals via cometabolism [22]. *Stenotrophomonas maltophilia* DT1 biotransforms tetracycline [23]. *Klebsiella* sp. SQY5 degrades tetracycline (TEC) [24]. Sulfamethoxazole is biodegraded by individual and mixed bacteria [25,26]. Antibiotic sulfanilamide biodegradation is performed by acclimated microbial populations [27]. Chlortetracycline can be used as the sole carbon and nitrogen source by the acclimated microbiota [28].

The aim of this study was to examine the degradation of antibiotics in sludge by antibiotic-degrading bacteria under aerobic and anaerobic conditions. The microbial communities involved in the aerobic and anaerobic degradation of antibiotics in sludge are revealed.

2. Materials and Methods

2.1. Chemicals

SMX, SDM, SMZ, TC, OTC, CTC, AMO (Table S1) of 99.0% purity and all other chemicals were purchased from Sigma Chemical Co. (St. Louis, MO, USA). Solvents were purchased from Mallinckrodt (Paris, KY, USA).

2.2. Sludge Sample and Medium

The sludge was a semisolid slurry (total solids 0.87 g/L, pH 6.7) that was produced as sewage sludge from the Dihus wastewater treatment plant in Taipei, Taiwan. Fresh sludge was used for antibiotic adaptation. The temperature of the sampling site was 30 °C. The medium used in aerobic experiments contained the following chemicals (mg/L): K_2HPO_4, 65.3; KH_2PO_4, 25.5; $Na_2HPO_4 \cdot 12 H_2O$, 133.8; NH_4Cl, 5.1; $CaCl_2$, 82.5; $MgSO_4 \cdot 7H_2O$, 67.5; and $FeCl_3 \cdot 6H_2O$, 0.75. The medium used in anaerobic experiments contained the following chemicals (mg/L): NH_4Cl, 2.7; $MgCl_2 \bullet 6H_2O$, 0.1; $CaCl_2 \bullet 2H_2O$, 0.1; $FeCl_2 \bullet 4H_2O$, 0.02; K_2HPO_4, 0.27; KH_2PO_4, 0.35 and resazurin, 0.001. The pH of the medium was adjusted to 7.0 using potassium hydroxide or nitric acid. Resazurin is an indicator, which exhibits red color under aerobic conditions and is colorless under anaerobic conditions. Titanium citrate of 0.9 mM was used as a reducing reagent. All anaerobic operations were performed in an anaerobic glove box.

2.3. Sludge Adaptation

Aerobic adaptation was performed by adding 1 mg/L CTC, SMX and AMO, simultaneously, to 1000 mL serum bottles containing 450 mL of aerobic medium, 50 mL sludge, and incubated on a rotary shaker (120 rpm) at 30 °C without light for 6 months. Anaerobic adaptation was performed by adding 1 mg/L CTC, SMX and AMO, simultaneously, to 1000 mL serum bottles containing 450 mL of anaerobic medium, 50 mL sludge and capped with butyl rubber stoppers and crimp seals, wrapped in aluminum foil, and then incubated without shaking at 30 °C without light for 6 months. All anaerobic operations were performed in an anaerobic glove box. In this paper, the sludge was referred to as antibiotic-adapted sludge.

2.4. Enrichment, Isolation, and Identification of Antibiotic-Degrading Bacteria

The enrichment procedure was performed using 5 mL of antibiotic-adapted sludge in a 125-mL serum bottle containing 45 mL of aerobic or anaerobic medium, with CTC, SMX and AMO of final concentration of 0.2 mg/L, simultaneously, and incubated at 30 °C. The second to the fourth enrichment transfers were amended with gradually increasing concentrations of CTC, SMX and AMO from 0.2 to

1 mg/L. An elevated concentration is commonly used for enrichment to ensure that the CTC, SMX and AMO degraders are selected. After the fourth enrichment, aerobic or anaerobic medium agar plates containing CTC, SMX and AMO (1 mg/L) were inoculated with 100 μL of the liquid part of the sludge by streaking to isolate pure strains of bacteria. The AnaeroPack system was used for anaerobic cultivation, including AnaeroPack®-Anaero and AnaeroPouch®-Anaero. All anaerobic operations were performed in the anaerobic glove box. To confirm that the bacterial strains were antibiotic degraders, degradation experiments were performed using 5 mL (10^6 CFU/mL) of bacterial culture and 45 mL of medium with 1 mg/L antibiotics on a shaker (120 rpm) at 30 °C in the dark. Samples were taken periodically to analyze residual CTC, SMX and AMO.

The 16S rRNA gene of the isolated bacterial strains were amplified by PCR with the 5′-primer F8 (5′-AGAGTTTGATCCTGGCTCAG-3′) and the 3′-primer R1510 (5′-GGTTACCTTGTTACGACTT-3′). The PCR parameters included initial denaturation at 94 °C for 10 min, followed by 35 cycles of 45 s at 94 °C, 1 min at 60 °C, and 1 min at 72 °C, with a final extension at 72 °C for 10 min. The PCR products were sequenced on an ABI Prism automatic sequencer. The 3′-end sequence was converted into the reverse complementary sequence. The overlapping parts of sequences from the 5′ and 3′-ends were identified with the Align Sequences Nucleotide BLAST tool at the National Center for Biotechnology Information (NCBI) website. Finally, the two sequences were assembled into a single contig sequence based on the overlapping sequences. The 16S rRNA sequences of the four isolates have been submitted to Genbank; accession number: A12: MT678104, B: MT678105, SANA: MT678106, SF1: MT678107. The 16S rRNA gene sequences of the four isolated bacterial strains were used to search the NCBI 16S rRNA database with the Basic Local Alignment Search Tool (BLASTn). The top five sequences (with the highest scores) in the Blast results for each 16S rRNA sequence were retrieved and used to construct the phylogenetic tree. Phylogenetic analysis was performed using Clustal X 2.0 with 1000 bootstrapping repetitions [29]. The neighbor-joining algorithm was used to construct the phylogenetic tree.

2.5. Settings of Batch and Readdition Experiments

Two sets of experiments (batch and continuous addition) were performed under aerobic and anaerobic conditions. In the batch experiments, the aerobic experiments were performed in 125 mL serum bottles with the following contents: 45 mL of medium, 5 g of sludge, 5 mL (10^6 CFU/mL) of bacterial culture, and 2 mg/L SAs (SMX, SDM, SMZ), TCs (TC, OTC, CTC) or AMO, and which were incubated on a rotary shaker (120 rpm) at 30 °C in the dark. The anaerobic experiments were performed in 125 mL serum bottles containing 45 mL of medium, 5 g of sludge, 5 mL (10^6 CFU/mL) of bacterial culture, and 2 mg/L SAs (SMX, SDM, SMZ), TCs (TC, OTC, CTC) or AMO, and were conducted in an anaerobic glove box. The bottles were capped with butyl rubber stoppers and crimp seals, wrapped in aluminum foil, and then incubated without shaking at 30 °C. Inoculated controls containing 45 mL of medium, 5 g of sludge, and 2 mg/L SAs (SMX, SDM, SMZ), TCs (TC, OTC, CTC) or AMO, and were incubated without degrading bacteria at 30 °C. Sterile controls containing 45 mL of medium and 5 g of sludge were autoclaved at 121 °C for 30 min. SAs (SMX, SDM, SMZ), TCs (TC, OTC, CTC) or AMO were added at 2 mg/L after autoclaving. The remaining percentage of the original antibiotic content in the sludge ranged from 98.5 to 96.8%, indicating that the aerobic and anaerobic antibiotic degradation observed in all of the following experiments was due to microbial activity. Each experiment was repeated three times.

For the repeated addition experiments, sludge samples were used by adding 2 mg/L CTC, SMX and AMO three times under aerobic or anaerobic conditions. Aerobic experiments were conducted using 1000 mL bottles containing 450 mL of aerobic medium, and 50 g of sludge, with or without 50 mL (10^6 CFU/mL) of mixed bacterial culture. Anaerobic experiments were filled using 1000 mL bottles containing 450 mL of anaerobic medium, and 50 g of sludge, with or without 50 mL (10^6 CFU/mL) of mixed bacterial culture. CTC, SMX and AMO was readded into each medium when antibiotics were decreased to an undetectable level. The aerobic experiments were aerated by an air diffuser and the mixtures were stirred. The anaerobic experiments were conducted in an anaerobic glove box;

capped with butyl rubber stoppers and crimp seals, wrapped in aluminum foil, and then incubated without shaking. The repeated addition experiments were performed at 30 °C.

2.6. Analytical Methods

SAs were extracted twice from the samples with an extraction solution of 10:3:1 water (containing 0.1% formic acid): acetonitrile:methanol. The extracts were analyzed using an Agilent 1260 HPLC system equipped with a 4.6 × 250 mm column (Zorbax Eclipse Plus C18, Agilent, Santa Clara, CA, USA) with photodiode array detector monitoring at 270 nm. The mobile phase was 30%:70% acetonitrile:water (containing 0.1% formic acid). TCs were extracted twice from the samples with an extraction solution of 4:1 water (containing 0.1% HCl): acetonitrile. The mobile phase consisted of acetonitrile and water (containing 0.1% HCl); AMO was extracted twice from the samples with methanol. TCs, and AMO were quantified with an Agilent 1260 HPLC equipped with a 4.6 × 100 mm column (Poroshell 120 EC-C18, Agilent) with a photodiode array detector monitoring at 270 nm and 229 nm. The recovery percentages for TC, OTC, CTC, SDM, SMX, SMZ, and AMO were 91.2%, 95.5%, 89.3%, 94.7%, 93.4%, 90.3% and 92.6%, respectively. The detection limit was 0.1 mg/L in all cases. The remaining percentage and removal percentage were computed using the following formula: remaining percentage [%] = (residue concentration/initial concentration) × 100; removal percentage [%] = [1 − (residue concentration/initial concentration) × 100]. The degradation data collected in this study fit well with first-order kinetics (i.e., $t = -\ln(C/C_0)/k$, where C_0 is the initial concentration, C is the concentration, t is the time period, $t_{1/2}$ is the half-life, and k is the degradation rate constant). The coefficient of determinations (R^2) ranged from 0.923 to 0.982.

2.7. Microbial Community Analysis

Total DNA was extracted from the sludge at the end of the readdition experiments using a PowerSoil DNA Isolation kit (QIAGEN, Venlo, Netherlands). The purified DNA was used as a template for the amplification of the 16S rRNA gene sequence containing the V5-V8 variable regions using a 5′-primer containing an Illumina adaptor and a 16S rRNA gene-specific sequence (5′-CCTACGGGNBGCASCAG-3′). The sequence of the 3′-primer contained an Illumina adaptor and a 16S rRNA gene-specific sequence (5′-GACTACNVGGGTATCTAATCC-3′). PCR was performed as previously described [30,31]. Next generation sequencing (NGS) was performed at the Genome Center of National Yang-Ming University, Taiwan, using the MiSeq platform (Illumina, Inc, San Diego, CA, USA.). A chimera check was used to analyze the 16S rRNA gene sequences to remove chimeras. The remaining sequences were subsequently analyzed using the classifier software of the Ribosomal Database Project (http://pyro.cme.msu.edu/) for phylogenetic assignment. Similarity (95%) was used for bacterial grouping. Xenobiotic biodegradation-associated bacteria were identified using the Kyoto Encyclopedia of Genes and Genomes (KEGG) database [32]. A cluster analysis of the bacterial community compositions was performed using the Heatmap function in the ComplexHeatmap package of R (www.r-project.org).

2.8. Scanning Electron Microscopy

Sludge samples were fixed in 2.5% glutaraldehyde in phosphate-buffered saline for 2 h at 4 °C and postfixed in 1% osmium tetroxide in the same buffer for 1 h. The samples were dehydrated in a graded ethyl alcohol series, critical point-dried, gold coated, and viewed under a JSM-5410 scanning electron microscope (JEOL Ltd., Tokyo, Japan).

3. Results and Discussion

3.1. Isolation and Identification of Antibiotic-Degrading Bacteria

Twelve and eight bacterial strains with the ability to use CTC, AMO and SMX as carbon sources under aerobic and anaerobic conditions were isolated from the sludge samples. The five aerobic isolates showing the greatest degrading capability were strains SF1, SF2, SF3, SF4 and A12. When the bacterial counts increased, the level of CTC, AMO and SMX decreased. The order of CTC, AMO and SMX degradation was strain SF1 > strain A12 > strain SF2 > strain SF3 > strain SF4. Strains SF1 and A12 exhibited the greatest aerobic degradation capability. The degradation of CTC, AMO and SMX by strain SF1 after 8 days of incubation was equal to 81.6, 89.1, and 95.9%, respectively (Table 1). The degradation of CTC, AMO and SMX by strain A12 after 8 days of incubation was equal to 69.4, 81.4, and 89.6%, respectively. The degradation of CTC, AMO and SMX by strains SF1 and A12 after 8 days of incubation was equal to 89.4, 93.4, and 99.3%, respectively. The order of the efficiency of the aerobic degradation of antibiotics was as follows: strains SF1 and A12 > strain SF1 > strain A12.

Table 1. Remaining percentages (%) of chlortetracycline (CTC), amoxicillin (AMO) and sulfamethoxazole (SMX) after incubation with antibiotic-degrading bacteria.

	Aerobic Condition [a]				Anaerobic Condition [b]			
	Medium	SF1	A12	SF1+A12	Medium	B	SANA	B + SANA
CTC	97.2 ± 3.70	18.4 ± 0.93	30.6 ± 1.48	10.6 ± 0.53	98.3 ± 4.33	18.4 ± 0.94	40.6 ± 2.13	10.4 ± 0.42
AMO	94.5 ± 2.91	10.9 ± 0.49	18.6 ± 0.88	6.6 ± 0.34	96.7 ± 5.24	10.9 ± 0.58	18.6 ± 0.90	6.4 ± 0.29
SMX	98.6 ± 5.12	4.1 ± 0.12	10.4 ± 0.41	0.7 ± 0.02	99.1 ± 6.78	4.9 ± 0.10	10.4 ± 0.44	0.5 ± 0.01

Note(s): [a] aerobic conditions for 8 days. [b] anaerobic conditions for 15 days. Data from three independent experiments are presented as the means ± SE.

Five anaerobic isolates exhibited the greatest antibiotic-degradation capability were strains B, SANA, SANB, SANC, SANC, and SAND. When the bacterial counts increased, the level of CTC, AMO and SMX decreased. The order of CTC, AMO and SMX degradation is strain B > strain SANA > strain SANB > strain SANC > strain SAND. Strains B and SANA exhibited the greatest antibiotic degradation capability under anaerobic conditions among the five tested isolates (B, SANA, SANB, SANC, SANC, and SAND). The degradation of CTC, AMO and SMX by strain B after 15 days of incubation was equal to 81.6, 89.1, and 95.1%, respectively (Table 1). The degradation of CTC, AMO and SMX by strain SANA after 15 days of incubation was equal to 59.4, 81.4, and 89.6%, respectively. The degradation of CTC, AMO and SMX by strain B and SANA after 15 days of incubation was equal to 89.6, 93.4, and 99.5%, respectively. The order of the efficiency of the anaerobic degradation of antibiotics was as follows: strains B and SANA > strain B > strain SANA. Yang et al. (2019) reported that the cocultures of strains M10 and M12 can enhance malachite green degradation in milkfish pond sediments. The addition of both bacterial strains M10 and M12 produced better results than each of the single cultures [31]. The aerobic stains SF1 and A12 and the anaerobic strains B and SANA were used in subsequent studies.

The colony morphology and scanning electron micrographs of the four bacterial strains are shown in Figure 1. All of them were Gram-negative, rod-shaped bacteria.

Phylogenetic analysis of strains A12, SF1, B and SANA, based on the 16S rRNA gene, is shown in Figure 2. The strains A12, SF1, B and SANA are closely related to *Pseudmonas pseudoalcaligenes* (99%), *Pseudmonas taiwanensis* (96%), *Bacillus flexus* (99%) and *Clostridium butyricum* (99%), respectively. *Pseudomonas* bacteria are widespread in various natural environments, such as soil, plants, animals, air and water. *Pseudmonas taiwanensis* is an aerobic Gram-negative, rod-shaped, motile, nonspore-forming bacterial strain isolated from soil [33]. *Pseudomonas pseudoalcaligenes* is an aerobic Gram-negative bacterium. It is able to use cyanide as a nitrogen source, and can be used for bioremediation [34]. *Bacillus flexus* is a Gram-positive, rod-shaped, endospore-forming bacterium. This bacterium may be isolated from feces and soil. *Bacillus flexus* has been shown to exert activity mainly against polyvinyl chloride additives and exhibits a low biodegradation rate of polyvinyl chloride polymers [35]. *Clostridium butyricum is* a strictly anaerobic endospore-forming Gram-positive bacillus. This bacterium has been studied for its efficiency in decolorizing various remazol reactive dyes [36].

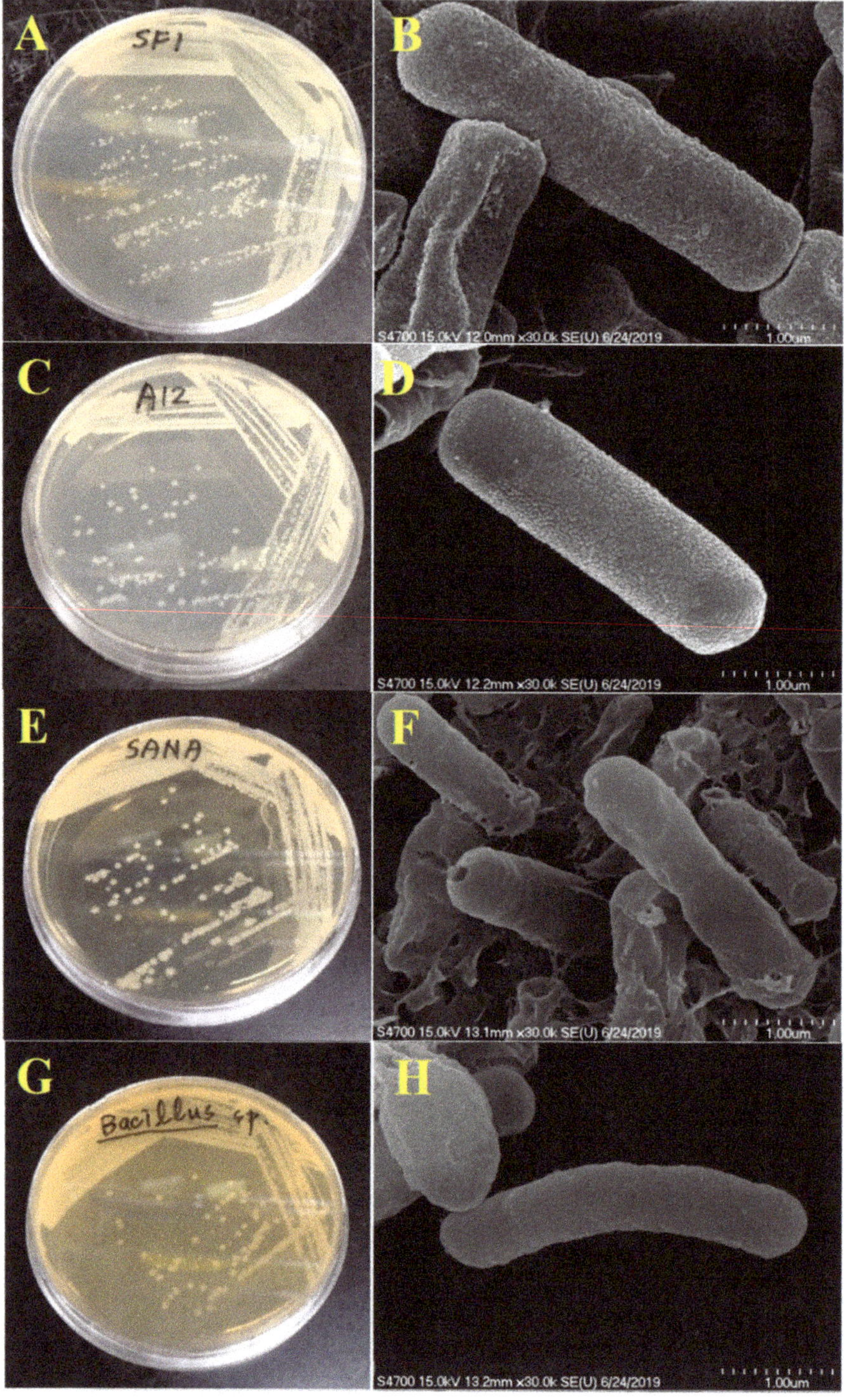

Figure 1. Colony morphology and scanning electron micrographs of isolated bacterial strains SF1 (Pseudmonas sp.) (**A**,**B**), A12 (Pseudmonas sp.) (**C**,**D**), SANA (Clostridium sp.) (**E**,**F**), and B (Bacillus sp.) (**G**,**H**).

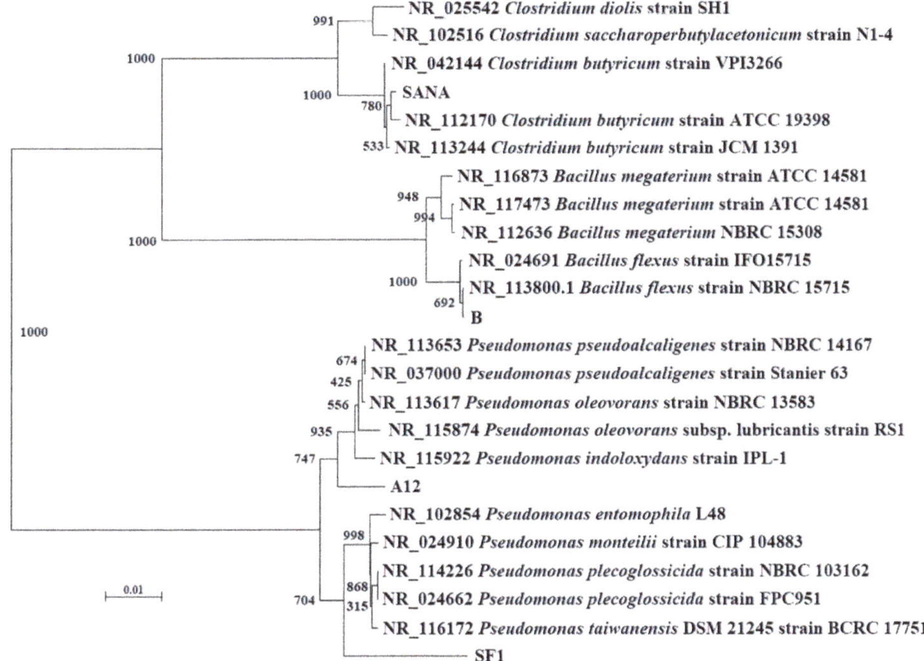

Figure 2. Phylogenetic analysis of the 16S rRNA genes of the four bacterial strains (A12, B, SF1, SANA). Bootstrapping values at branch points indicate the number of times that the same branch was observed out of 1000 repeats of the phylogenetic reconstruction.

3.2. Degradation of Antibiotics in Sludge with Isolated Bacteria

The antibiotic concentrations in the sterile controls were first examined at the end of the 10th or 15th day incubation periods, under aerobic or anaerobic conditions, respectively. The remaining percentages of antibiotics ranged from 91.7 to 95.3%. Therefore, it was concluded that the antibiotic degradation that occurred in all of the following experiments was due to microbial action.

The aerobic and anaerobic degradation of the antibiotics in sludge containing the isolated bacterial strains is shown in Figures 3–5. The degradation of the antibiotics was increased in sludge containing the isolated bacterial strains. As shown in Figures 3A, 4A and 5A, the aerobic degradation half-lives of OTC, TC, CTC, SDM, SMX, SMZ and AMO were 3.0, 4.6, 2.3, 2.5, 3.1, 1.7 and 2.0 days, respectively. The anaerobic degradation half-lives of OTC, TC, CTC, SDM, SMX, SMZ and AMO were 4.4, 7.0, 3.4, 4.1, 6.8, 2.4 and 3.0 days, respectively (Figures 3B, 4B and 5B). The order of the degradation rates of the antibiotics was CTC > OTC > TC> AMO > SMX > SDM > SMZ.

The order of SA degradation rates in sludge was SMX > SDM > SMZ. Similar trends were observed in a study by Yang et al. (2016) in which fungal enzymes were used to increase the degradation of SAs in sludge [37]. The order of the TC degradation rates in sludge was CTC > OTC > TC. Similar results were reported by Suda et al. (2012) from a study in which TC antibiotics were treated with the laccase enzyme in the presence of 1-hydroxybenzotriazole [38]. The order of the antibiotic degradation rates was TCs > AMO > SAs. Antibiotics with lower molecular weights are easier to degrade than those with higher molecular weights. Larger functional groups may hinder degradation by affecting the interactions between target compounds and the bacteria or enzymes [39]. Moreover, the degradation of compounds with complex structures may require more reaction steps.

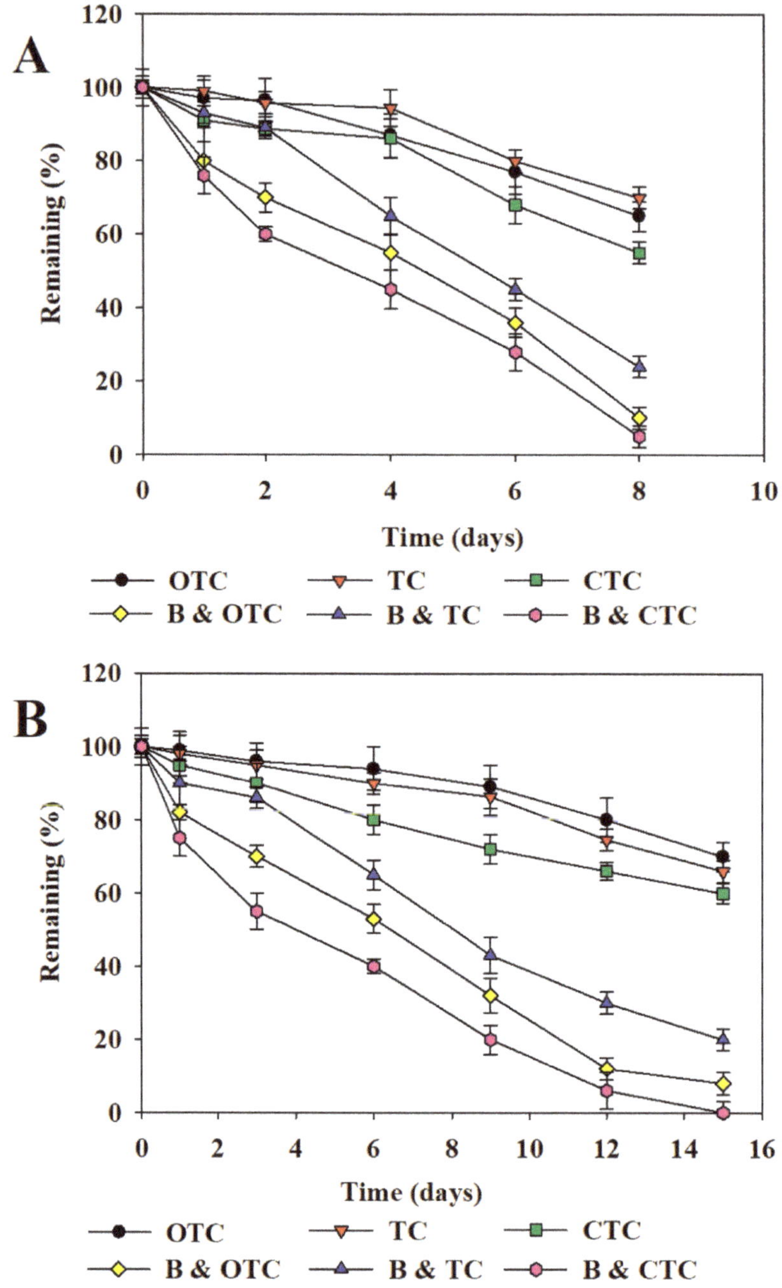

Figure 3. Degradation of tetracyclines with and without degrading bacteria in sludge. Aerobic (**A**) and anaerobic (**B**) degradation of TCs. TC: tetracycline, CTC: chlortetracycline, OTC: oxytetracycline. B: isolated antibiotic-degrading bacteria. Data from three independent experiments are presented as the means ± SE.

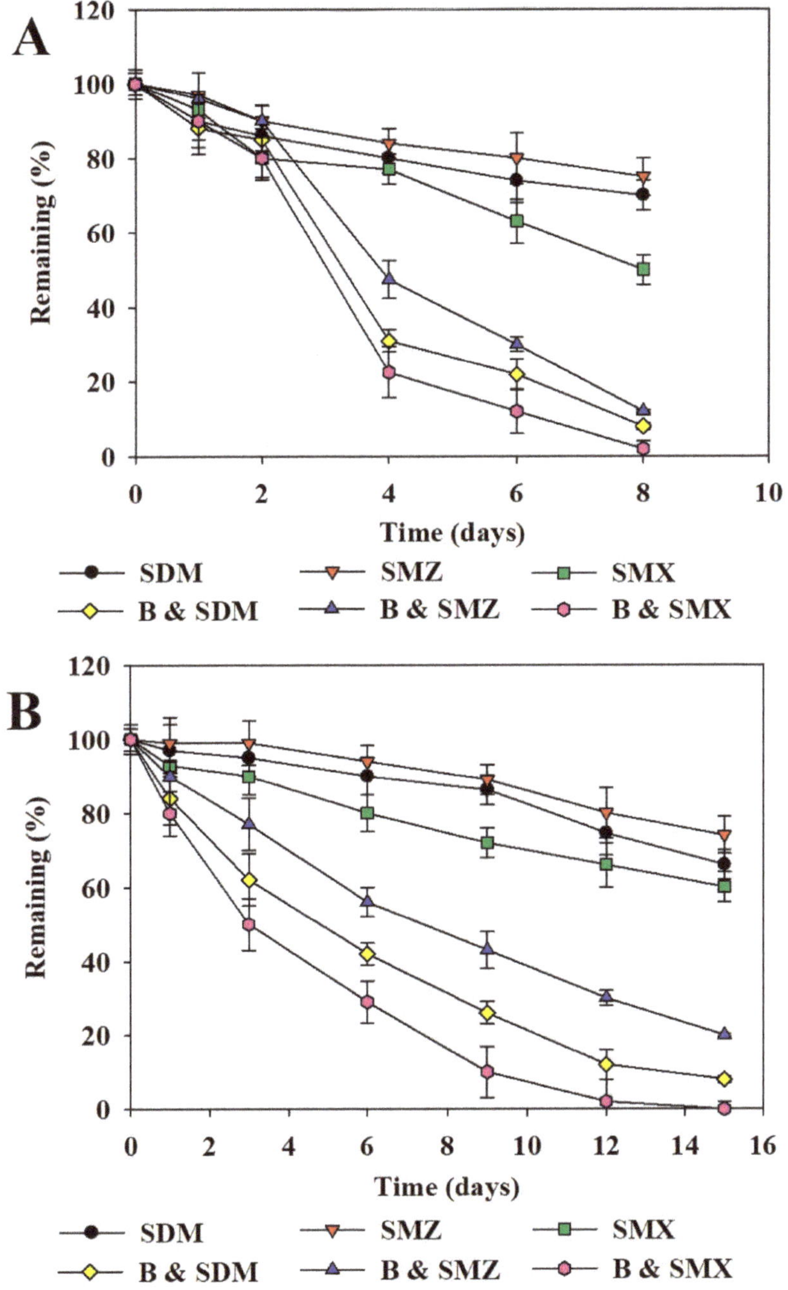

Figure 4. Degradation of sulfonamides with and without degrading bacteria in sludge. Aerobic (**A**) and anaerobic (**B**) degradation of SAs. SMZ: sulfamethazine, SMX: sulfamethoxazole, SDM: sulfadimethoxine. B: isolated antibiotic-degrading bacteria. Data from three independent experiments are presented as the means ± SE.

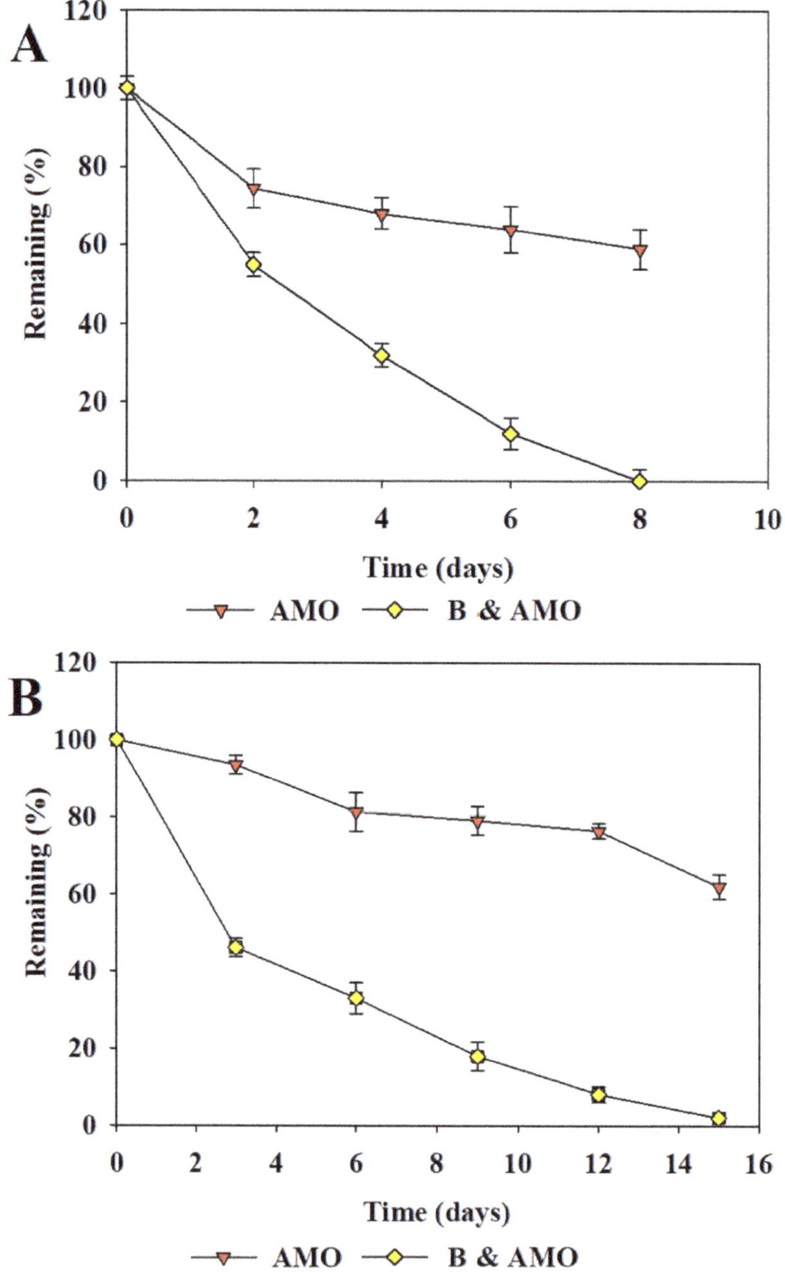

Figure 5. Degradation of amoxicillin with and without degrading bacteria in sludge. Aerobic (**A**) and anaerobic (**B**) degradation of AMO. AMO: amoxicillin. B: isolated antibiotic-degrading bacteria. Data from three independent experiments are presented as the means ± SE.

3.3. Repeated Addition of Antibiotics and Their Degradation in Sludge

To test the long-term degradation ability of aerobic strains SF1 and A12 and anaerobic strains B and SANA in sludge, experiments involving the repeated addition of antibiotics were performed (Table 2).

Table 2. The removal rates (%) after the first (1st), second (2nd), and third (3rd) additions of antibiotics.

		CTC	AMO	SMX
1st	Sludge [a]	35.6 ± 1.6	41.6 ± 1.8	46.3 ± 1.5
	Sludge & SF1 & A12 [a]	82.6 ± 2.5	88.2 ± 2.9	90.6 ± 3.8
	Sludge [b]	33.5 ± 1.2	37.4 ± 1.1	42.2 ± 1.6
	Sludge & B & SANA[b]	85.2 ± 2.4	88.1 ± 2.1	92.3 ± 3.1
2nd	Sludge [a]	43.7 ± 1.5	45.3 ± 1.3	49.1 ± 1.4
	Sludge & SF1 & A12 [a]	88.6 ± 2.5	90.2 ± 2.9	92.5 ± 2.9
	Sludge [b]	41.2 ± 1.5	43.7 ± 1.3	46.4 ± 1.2
	Sludge & B & SANA [b]	90.7 ± 3.3	91.1 ± 3.1	94.3 ± 3.5
3rd	Sludge [a]	46.1 ± 1.3	50.4 ± 1.3	52.5 ± 1.1
	Sludge & SF1 & A12 [a]	90.3 ± 3.2	95.3 ± 4.1	97.5 ± 3.9
	Sludge [b]	42.7 ± 1.1	46.4 ± 1.3	50.4 ± 1.2
	Sludge & B & SANA [b]	92.4 ± 4.2	96.4 ± 4.4	98.1 ± 4.7

Note(s): [a] aerobic conditions for 8 days. [b] anaerobic conditions for 15 days. Data from three independent experiments are presented as the means ± SE.

In the first addition experiments, in the sludge under aerobic conditions, 46.3, 41.6 and 35.6% of SMX, AMO and CTC initially present at 2 mg/L was degraded, respectively. In the sludge under anaerobic conditions, 42.2, 37.4 and 33.5% of SMX, AMO and CTC initially present at 2 mg/L was degraded, respectively. In the sludge under aerobic conditions containing aerobic strains SF1 and A12, 90.6, 88.2 and 82.6% of SMX, AMO and CTC initially present at 2 mg/L was degraded, respectively. In the sludge under anaerobic conditions containing anaerobic strains B and SANA, 92.3, 88.1 and 85.2% of SMX, AMO and CTC initially present at 2 mg/L was degraded, respectively. In the third addition experiments, in the sludge under aerobic conditions, 52.5, 50.4 and 46.1% of SMX, AMO and CTC initially present at 2 mg/L was degraded, respectively. In the sludge under anaerobic conditions, 50.4, 46.4 and 42.7% of SMX, AMO and CTC initially present at 2 mg/L was degraded, respectively. In the sludge under aerobic conditions containing aerobic strains SF1 and A12, 97.5, 95.3 and 90.3% of SMX, AMO and CTC initially present at 2 mg/L was degraded, respectively. In the sludge under anaerobic conditions containing anaerobic strains B and SANA, 98.1, 96.4 and 92.4% of SMX, AMO and CTC initially present at 2 mg/L was degraded, respectively. These results suggest that the degradation rate increased with the repeated addition of SMX, AMO and CTC. The repeated addition of antibiotics might increase the abundance of antibiotic-degrading microbes. These findings are similar to the results of our previous study of SA degradation in sludge [38].

3.4. Microbial Communities in the Repeated Addition Experiments

To obtain deeper insight into the conditions in the repeated addition experiments, the microbial communities in the repeated addition experiments were analyzed. As shown in Figure 6, the five experimental conditions (original sludge, sludge in aerobic conditions after the 3rd addition, sludge and bacteria in aerobic conditions after the 3rd addition, sludge in anaerobic conditions after the 3rd addition, sludge and bacteria in anaerobic conditions after the 3rd addition) exhibited differences in the composition of the major microbial communities (46, 45, 24, 33 and 26 microbial genera). The lists of the five groups of major microbial genera in the five experimental conditions are listed in Table S2.

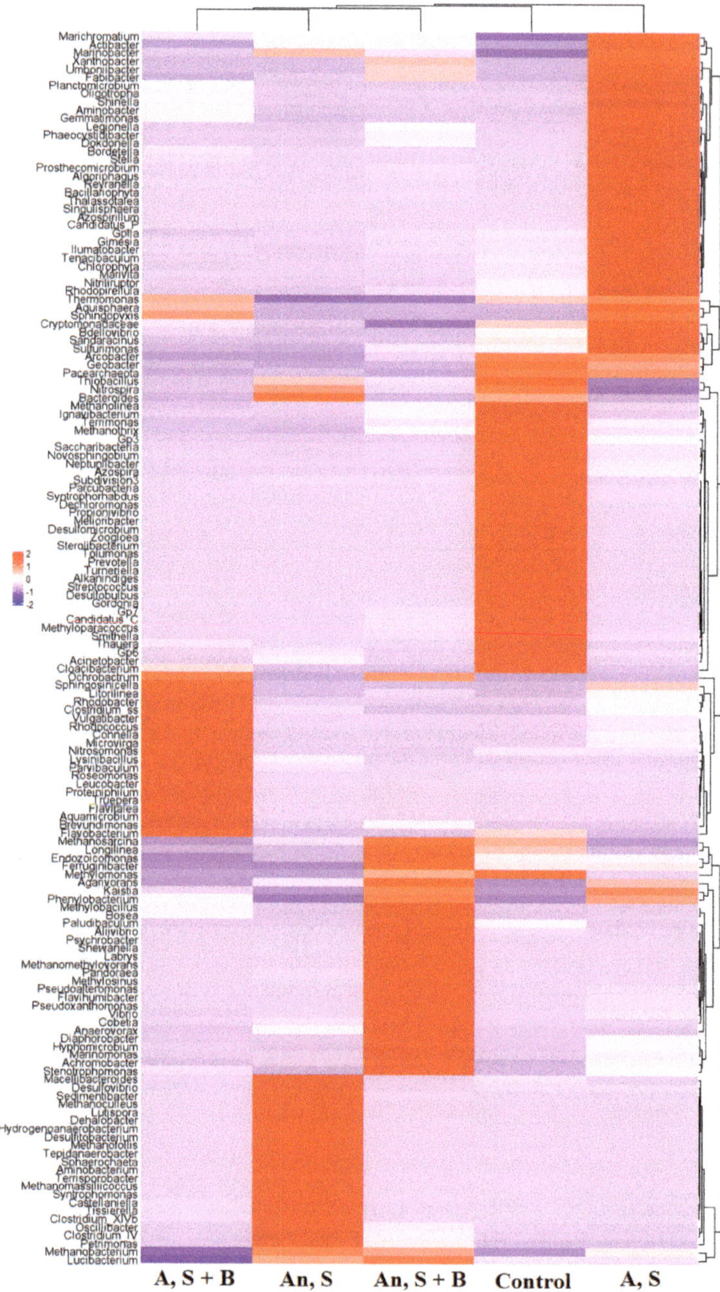

Figure 6. Major microbial communities (genus level) in sludge after the 3rd repeated addition experiment. Control: original sludge. A, S: aerobic conditions, sludge only. A, S + B: aerobic conditions, sludge + bacteria. An, S: anaerobic conditions, sludge only. An, S + B: anaerobic conditions, sludge + bacteria. The major microbial genera (red color in heatmap) under the five experimental conditions are listed in Table S2.

For validation, a list of xenobiotic biodegradation-associated bacteria/archaea in the KEGG database was used to identify bacteria/archaea with xenobiotic degradation ability in the repeated addition experiments. The identified bacterial/archaeal genera were used as key words to perform text mining using the NCBI PubMed database. Under aerobic conditions, 25 of the 63 identified bacterial/archaeal genera have been reported to be antibiotic-degrading bacteria/archaea (Figure 7A). Under anaerobic conditions, 25 of the 63 identified bacterial/archaeal genera have been reported to be antibiotic-degrading bacteria/archaea (Figure 7B).

In total, 24 reported antibiotic-degrading bacterial genera (*Achromobacter, Acidovorax, Acinetobacter, Alcaligenes, Bacillus, Burkholderia, Castellaniella, Comamonas, Corynebacterium, Cupriavidus, Dechloromonas, Geobacter, Gordonia, Klebsiella, Mycobacterium, Novosphingobium, Pandoraea, Pseudomonas, Rhodococcus, Sphingomonas, Thauera, Treponema, Vibrio and Xanthobacter*) are common to the aerobic and anaerobic groups. These results suggest that the 24 bacterial genera may be the major antibiotic-degrading bacteria in sludge. The PubMed ID, title and abstract of these reports are collected and summarized in Table S3.

The distribution of the number of major microbial genera exhibiting different aromatic compound degradation pathways is shown in Figure 8. Most of the microbial genera were associated with six reaction modules: M00548 (benzene degradation, benzene ≥ catechol), M00551 (benzoate degradation, benzoate ≥ catechol/methylbenzoate ≥ methylcatechol), M00568 (catechol ortho-cleavage, catechol ≥ 3-oxoadipate), M00569 (catechol meta-cleavage, catechol ≥ acetyl-CoA/4-methylcatechol ≥ propanoyl-CoA), M00623 (phthalate degradation, phthalate ≥ protocatechuate) and M00638 (salicylate degradation, salicylate ≥ gentisate). Moreover, more microbial genera associated with aromatic compound degradation were identified in the sludge with isolated antibiotic-degrading bacteria than in the sludge without the bacteria. These results provide explanations for the increased antibiotic-degrading effects associated with the addition of isolated antibiotic-degrading bacteria in sludge observed in Figures 3–5 and Table 2.

Yin et al. (2020) identified a TC-degrading strain, TR5, that could degrade TC quickly (~90% within 36 h) when the initial TC concentration was 200 mg/L [40]. Its efficiency is much higher than those of the TC-degrading bacteria identified in this study. However, TR5 was identified as *Klebsiella* pneumoniae according to 16S rRNA gene sequencing and biochemical properties. Great care must be taken in the application of pathogenic bacteria in wastewater treatment. Wu et al. (2020) identified two TC-degrading strains, *Raoultella* sp. XY-1 and *Pandoraea* sp. XY-2, which degraded 81.72% TC within 12 days in lysogeny broth (LB) medium [41]. The bacterial strains identified in this study exhibit a higher TC-degrading efficiency than *Raoultella* sp. XY-1 and *Pandoraea* sp. XY-2. Wen et al. reported that 95% of the doxycycline (50 mg/L) was degraded by the recombinant strain *Escherichia coli* ETD-1, with tetX, within 48 h [42]. Although the doxycycline-degrading ability of recombinant *Escherichia coli* is very significant, great care must be taken in the application of microorganisms carrying recombinant genes in the environment. Sodhi et al. (2020) identified the AMO-degrading bacterium, *Alcaligenes* sp. MMA, which was able to remove up to 84% of amoxicillin in 14 days in M9 minimal media [43]. The bacterial strains identified in this study exhibit a higher AMO-degrading efficiency than *Alcaligenes* sp. MMA. Liang et al. (2019) reported that *Achromobacter* sp. JL9 was able to utilize SMX as its sole nitrogen source for growth, with an SMX biodegradation efficiency of 63.10% [44]. Moreover, Nguyen et al. (2019) showed that SMX and sodium acetate could be cometabolized as carbon sources, with the highest removal efficiencies of 82.44%, 80.2%, and 79.45% for NH_4^+-N, NO_3^--N, and SMX, respectively [45]. The bacterial strains identified in this study exhibit a higher SMX-degrading efficiency than *Achromobacter* sp. Wang et al. (2018) demonstrated that *Acinetobacter* sp. could mineralize 98.8% of SMX, but only 17.5% and 20.5% of sulfadiazine and SMZ, respectively [46]. The bacterial strains identified in this study could degrade 98%, 91% and 88% SMX, SDM and SMZ, respectively. Overall, the antibiotic-degrading efficiency of the bacterial strains identified in this study is very good. Moreover, these bacterial strains exhibit the ability to degrade multiple antibiotics.

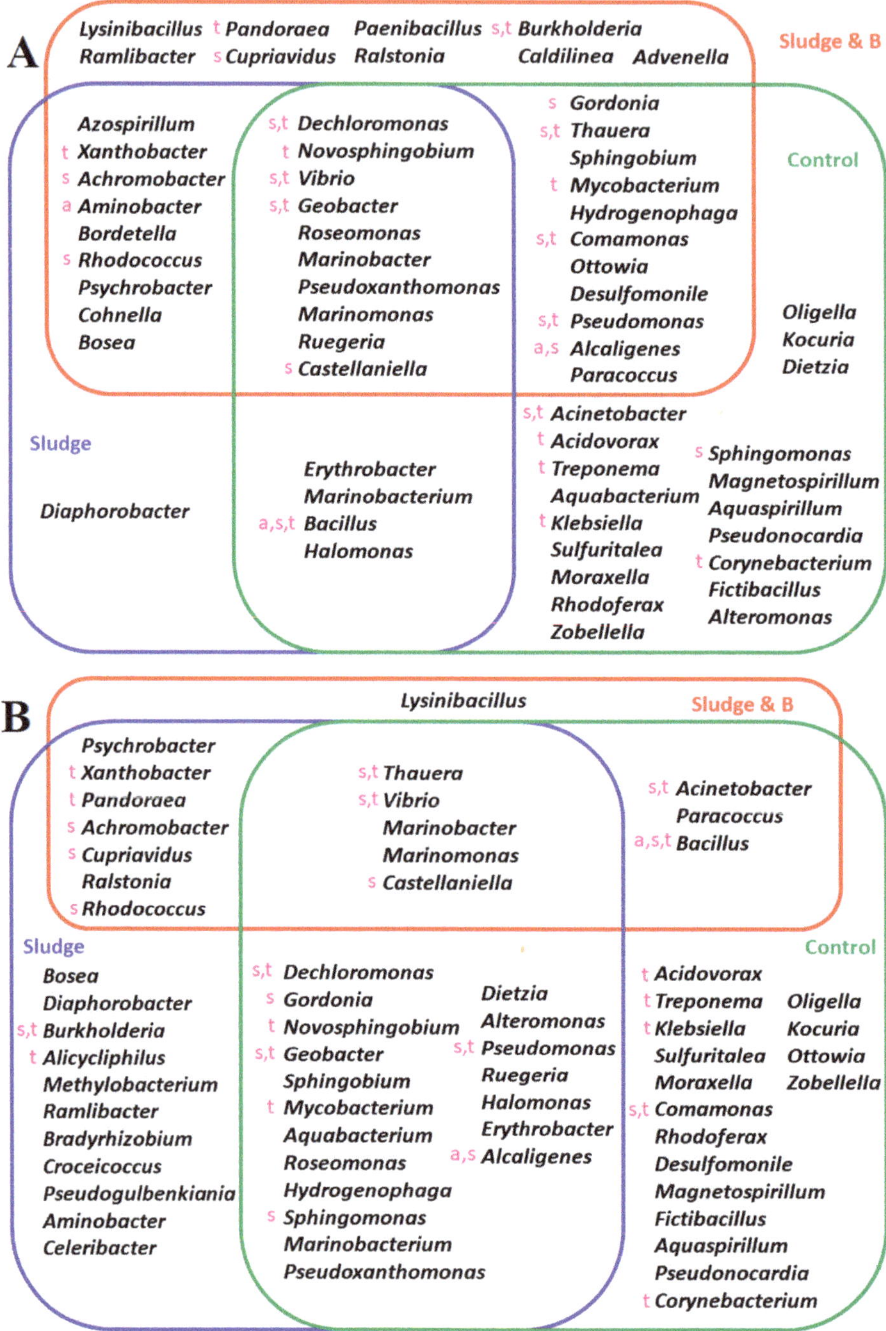

Figure 7. Xenobiotic biodegradation-associated bacteria after the 3rd repeated addition experiment under aerobic (**A**) and anaerobic (**B**) conditions. Control: original sludge. Sludge and B: sludge containing the isolated bacterial strains. a: amoxicillin-degrading bacteria. s: sulfonamide-degrading bacteria. t: tetracycline-degrading bacteria.

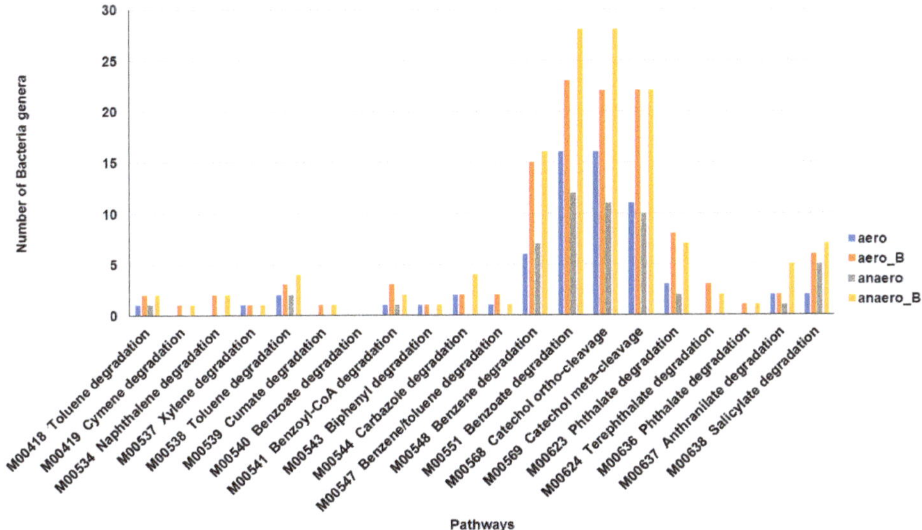

Figure 8. Xenobiotic biodegradation-associated microbial genera after the 3rd repeated addition experiment under aerobic and anaerobic conditions. Mxxxxx are Kyoto Encyclopedia of Genes and Genomes (KEGG) database module IDs. aero: sludge under aerobic conditions. anaero: sludge under anaerobic conditions. aero_B: sludge with antibiotic-degrading bacteria under aerobic conditions. anaero: sludge with antibiotic-degrading bacteria under anaerobic conditions.

4. Conclusion

The removal efficiency of antibiotics in wastewater sludge was enhanced using the bacterial strains isolated in this study. Batch experiments indicated that the combination of two bacterial strains enhanced the biodegradation of antibiotics in sludge to a greater extent than the application of a single bacterial strain under both aerobic and anaerobic conditions. The efficiency was comparable between the aerobic and anaerobic settings. However, anaerobic biodegradation required a longer incubation time (by 2-fold) than aerobic biodegradation. The results of repeated addition experiments indicated that the efficiency of the biodegradation of antibiotics using bacterial strains could be maintained for at least three degradation runs. The added antibiotic-degrading bacteria not only degraded antibiotics directly but also changed the composition of the microbial community in the sludge. As a consequence, the composition of the xenobiotic degradation pathways and the antibiotic degradation capacity were changed.

Supplementary Materials: The following are available online at http://www.mdpi.com/2073-4441/12/8/2147/s1, Table S1: Chemicals structures of target compounds, Table S2: five groups of major microbial genera in the five experimental conditions, Table S3: The PubMed ID, title and abstract of iterature reported antibiotics-degrading bacterial genera.

Author Contributions: Formal analysis, C.L.; Funding acquisition, B.-V.C.; Methodology, C.-W.Y. and C.L.; Project administration, B.-V.C.; Visualization, C.-W.Y.; Writing—original draft, C.-W.Y. and B.-V.C.; Writing—review and editing, C.-W.Y. and B.-V.C. All authors have read and agreed to the published version of the manuscript.

Funding: This research was supported by the Ministry of Science and Technology, Taiwan (grant no. MOST 107-2313-B-031-002).

Conflicts of Interest: The authors declare that they have no conflict of interest.

References

1. Salgado, R.; Noronha, J.P.; Oehmen, A.; Carvalho, G.; Reis, M.A. Analysis of 65 pharmaceuticals and personal care products in 5 wastewater treatment plants in Portugal using a simplified analytical methodology. *Water Sci. Technol.* **2010**, *62*, 2862–2871. [CrossRef]
2. Barber, L.B.; Keefe, S.H.; LeBlanc, D.R.; Bradley, P.M.; Chapelle, F.H.; Meyer, M.T.; Loftin, K.A.; Kolpin, D.W.; Rubio, F. Fate of sulfamethoxazole, 4- nonylphenol, and 17ß-estradiol in groundwater contaminated by wastewater treatment plant effluent. *Environ. Sci. Technol.* **2009**, *43*, 4843–4850. [CrossRef]
3. Qiao, M.; Ying, G.G.; Singer, A.C.; Zhu, Y.G. Review of antibiotic resistance in China and its environment. *Environ. Int.* **2018**, *110*, 160–172. [CrossRef]
4. Gzyl, K.E.; Wieden, H.J. Tetracycline does not directly inhibit the function of bacterial elongation factor Tu. *PLoS ONE* **2017**, *12*, e0178523. [CrossRef]
5. Sukul, P.; Spiteller, M. Sulfonamides in the environment as veterinary drugs. In *Reviews of Environmental Contamination and Toxicology*; Springer: New York, NY, USA, 2006; Volume 187, pp. 67–101.
6. Gobel, A.; Thomsen, A.; McArdell, C.S.; Joss, A.; Giger, W. Occurrence and sorption behavior of sulfonamides, macrolides, and trimethoprim in activated sludge treatment. *Environ. Sci. Technol.* **2005**, *39*, 3981–3989. [CrossRef] [PubMed]
7. Nieto, A.; Borrull, F.; Pocurull, E.; Marce, R.M. Occurrence of pharmaceuticals and hormones in sewage sludge. *Environ. Toxicol. Chem.* **2010**, *29*, 1484–1489. [CrossRef]
8. Adamek, E.; Baran, W.; Sobczak, A. Assessment of the biodegradability of selected sulfa drugs in two polluted rivers in Poland: Effects of seasonal variations, accidental contamination, turbidity and salinity. *J. Hazard Mater.* **2016**, *313*, 147–158. [CrossRef] [PubMed]
9. Baghapour, M.A.; Shirdarreh, M.R.; Faramarzian, M. Amoxicillin removal from aqueous solutions using submerged biological aerated filter. *Desalin. Water Treat.* **2014**, *54*, 790–801. [CrossRef]
10. Gavrilescu, M.; Demnerova, K.; Aamand, J.; Agathos, S.; Fava, F. Emerging pollutants in the environment: Present and future challenges in biomonitoring, ecological risks and bioremediation. *New Biotechnol.* **2015**, *32*, 147–156. [CrossRef] [PubMed]
11. Michael-Kordatou, I.; Karaolia, P.; Fatta-Kassinos, D. The role of operating parameters and oxidative damage mechanisms of advanced chemical oxidation processes in the combat against antibiotic-resistant bacteria and resistance genes present in urban wastewater. *Water Res.* **2018**, *129*, 208–230. [PubMed]
12. Chang, B.V.; Fan, S.N.; Tsai, Y.C.; Chung, Y.L.; Tu, P.X.; Yang, C.W. Removal of emerging contaminants using spent mushroom compost. *Sci. Total Environ.* **2018**, *634*, 922–933. [CrossRef] [PubMed]
13. Sun, C.; Li, W.; Chen, Z.; Qin, W.; Wen, X. Responses of antibiotics, antibiotic resistance genes, and mobile genetic elements in sewage sludge to thermal hydrolysis pre-treatment and various anaerobic digestion conditions. *Environ. Int.* **2019**, *133*, 105156. [CrossRef] [PubMed]
14. He, Z.; Wei, Z.; Zhao, Y.; Zhang, D.; Pan, X. Enhanced performance of tetracycline treatment in wastewater using aerobic granular sludge with in-situ generated biogenic manganese oxides. *Sci. Total Environ.* **2020**, *735*, 139533. [CrossRef] [PubMed]
15. Du, B.; Yang, Q.; Li, X.; Yuan, W.; Chen, Y.; Wang, R. Impacts of long-term exposure to tetracycline and sulfamethoxazole on the sludge granules in an anoxic-aerobic wastewater treatment system. *Sci. Total Environ.* **2019**, *684*, 67–77. [CrossRef]
16. Krzeminski, P.; Tomei, M.C.; Karaolia, P.; Langenhoff, A.; Almeida, C.M.; Felis, E.; Gritten, F.; Andersen, H.R.; Fernandes, T.; Manaia, C.M.; et al. Performance of secondary wastewater treatment methods for the removal of contaminants of emerging concern implicated in crop uptake and antibiotic resistance spread: A review. *Sci. Total Environ.* **2019**, *648*, 1052–1081. [CrossRef]
17. Bisognin, R.P.; Wolff, D.B.; Carissimi, E.; Prestes, O.D.; Zanella, R. Occurrence and fate of pharmaceuticals in effluent and sludge from a wastewater treatment plant in Brazil. *Environ. Technol.* **2019**, *2019*, 1–12. [CrossRef]
18. Ezzariai, A.; Hafidi, M.; Khadra, A.; Aemig, Q.; El Fels, L.; Barret, M.; Merlina, G.; Patureau, D.; Pinelli, E. Human and veterinary antibiotics during composting of sludge or manure: Global perspectives on persistence, degradation, and resistance genes. *J. Hazard Mater.* **2018**, *359*, 465–481. [CrossRef]
19. Boonnorat, J.; Kanyatrakul, A.; Prakhongsak, A.; Honda, R.; Panichnumsin, P.; Boonapatcharoen, N. Effect of hydraulic retention time on micropollutant biodegradation in activated sludge system augmented with acclimatized sludge treating low-micropollutants wastewater. *Chemosphere* **2019**, *230*, 606–615. [CrossRef]

20. Nas, B.; Argun, M.E.; Dolu, T.; Ateş, H.; Yel, E.; Koyuncu, S.; Dinç, S.; Kara, M. Occurrence, loadings and removal of EU-priority polycyclic aromatic hydrocarbons (PAHs) in wastewater and sludge by advanced biological treatment, stabilization pond and constructed wetland. *J. Environ. Manag.* **2020**, *268*, 110580. [CrossRef]
21. Kamaz, M.; Wickramasinghe, S.R.; Eswaranandam, S.; Zhang, W.; Jones, S.M.; Watts, M.J.; Qian, X. Investigation into micropollutant removal from wastewaters by a membrane bioreactor. *Int. J. Environ. Res. Public Health* **2019**, *16*, 1363. [CrossRef]
22. Gauthier, H.; Yargeau, V.; Cooper, D.G. Biodegradation of pharmaceuticals by *Rhodococcus rhodochrous* and *Aspergillus niger* by co-metabolism. *Sci. Total Environ.* **2010**, *408*, 1701–1706. [CrossRef] [PubMed]
23. Leng, Y.; Bao, J.; Chang, G.; Zheng, H.; Li, X.; Du, J.; Snow, D. Biotransformation of tetracycline by a novel bacterial strain *Stenotrophomonas maltophilia* DT1. *J. Hazard. Mater.* **2016**, *318*, 125–133. [CrossRef] [PubMed]
24. Shao, S.; Hu, Y.; Cheng, J.; Chen, Y. Biodegradation mechanism of tetracycline (TEC) by strain *Klebsiella* sp. SQY5 as revealed through products analysis and genomics. *Ecotoxicol. Environ. Saf.* **2019**, *85*, 109676–109683. [CrossRef]
25. Larcher, S.; Yargeau, V. Biodegradation of sulfamethoxazole by individual and mixed bacteria. *Appl. Microbiol. Biotechnol.* **2011**, *91*, 211–218. [CrossRef] [PubMed]
26. Wang, J.; Wang, S. Microbial degradation of sulfamethoxazole in the environment. *Appl. Microbiol. Biotechnol.* **2018**, *102*, 3573–3582. [CrossRef] [PubMed]
27. Liao, X.; Li, B.; Zou, R.; Xie, S.; Yuan, B. Antibiotic sulfanilamide biodegradation by acclimated microbial populations. *Appl. Microbiol. Biotechnol.* **2016**, *100*, 2439–2447. [CrossRef]
28. Liao, X.; Zou, R.; Li, B.; Tong, T.; Xie, S.; Yuan, B. Biodegradation of chlortetracycline by acclimated microbiota. *Process Saf. Environ.* **2017**, *109*, 11–17. [CrossRef]
29. Larkin, M.A.; Blackshields, G.; Brown, N.P.; Chenna, R.; McGettigan, P.A.; McWilliam, H.; Valentin, F.; Wallace, I.M.; Wilm, A.; Lopez, R.; et al. Clustal W and Clustal X version 2.0. *Bioinformatics* **2007**, *23*, 2947–2948. [CrossRef]
30. Chang, B.V.; Chang, Y.T.; Chao, W.L.; Yeh, S.L.; Kuo, D.L.; Yang, C.W. Effects of sulfamethoxazole and sulfamethoxazole-degrading bacteria on water quality and microbial communities in milkfish ponds. *Environ. Pollut.* **2019**, *252*, 305–316. [CrossRef]
31. Yang, C.W.; Chao, W.L.; Hsieh, C.Y.; Chang, B.V. Biodegradation of malachite green in milkfish pond sediments. *Sustainability* **2019**, *11*, 4179. [CrossRef]
32. Kanehisa, M.; Furumichi, M.; Tanabe, M.; Sato, Y.; Morishima, K. KEGG: New perspectives on genomes, pathways, diseases and drugs. *Nucleic Acids Res.* **2017**, *45*, D353–D361. [CrossRef] [PubMed]
33. Wang, L.T.; Tai, C.J.; Wu, Y.C.; Chen, Y.B.; Lee, F.L.; Wang, S.L. *Pseudomonas taiwanensis* sp. nov., isolated from soil. *Int. J. Syst. Evol. Micr.* **2010**, *60*, 2094–2098. [CrossRef] [PubMed]
34. Huertas, M.J.; Luque-Almagro, V.M.; Martinez-Luque, M.; Blasco, R.; Moreno-Vivian, C.; Castillo, F.; Roldan, M.D. Cyanide metabolism of *Pseudomonas pseudoalcaligenes* CECT5344: Role of siderophores. *Biochem. Soc. Trans.* **2006**, *34*, 152–155. [CrossRef] [PubMed]
35. Giacomucci, L.; Raddadi, N.; Soccio, M.; Lotti, N.; Fava, F. Polyvinyl chloride biodegradation by *Pseudomonas citronellolis* and *Bacillus flexus*. *New Biotechnol.* **2019**, *52*, 35–41. [CrossRef] [PubMed]
36. Ekambaram, S.P.; Perumal, S.S.; Annamalai, U. Decolorization and biodegradation of remazol reactive dyes by *Clostridium* species. *3 Biotech.* **2016**, *6*, 20. [CrossRef]
37. Yang, C.W.; Hsiao, W.C.; Chang, B.V. Biodegradation of sulfonamide antibiotics in sludge. *Chemosphere* **2016**, *150*, 559–565. [CrossRef]
38. Suda, T.; Hata, T.; Kawai, S.; Okamura, H.; Nishida, T. Treatment of tetracycline antibiotics by laccase in the presence of 1-hydroxybenzotriazole. *Bioresour. Technol.* **2012**, *103*, 498–501. [CrossRef]
39. Pepper, L.L.; Gerba, C.P.; Gentry, T.J. Environmental microbiology. In *Microorganisms and Organic Pollutant*; Maier, M.M., Gentry, T.J., Eds.; Elsevier Inc.: Berkeley, CA, USA, 2015; pp. 377–413.
40. Yin, Z.; Xia, D.; Shen, M.; Zhu, D.; Cai, H.; Wu, M.; Zhu, Q.; Kang, Y. Tetracycline degradation by *Klebsiella* sp. strain TR5: Proposed degradation pathway and possible genes involved. *Chemosphere* **2020**, *253*, 126729. [CrossRef]
41. Wu, X.; Gu, Y.; Wu, X.; Zhou, X.; Zhou, H.; Amanze, C.; Shen, L.; Zeng, W. Construction of a Tetracycline Degrading Bacterial Consortium and Its Application Evaluation in Laboratory-Scale Soil Remediation. *Microorganisms* **2020**, *8*, 292. [CrossRef]

42. Wen, X.; Huang, J.; Cao, J.; Xu, J.; Mi, J.; Wang, Y.; Ma, B.; Zou, Y.; Liao, X.; Liang, J.B.; et al. Heterologous expression of the tetracycline resistance gene *tetX* to enhance degradability and safety in doxycycline degradation. *Ecotoxicol. Environ. Saf.* **2020**, *191*, 110214. [CrossRef]
43. Sodhi, K.K.; Kumar, M.; Singh, D.K. Potential application in amoxicillin removal of *Alcaligenes* sp. MMA and enzymatic studies through molecular docking. *Arch. Microbiol.* **2020**. [CrossRef] [PubMed]
44. Liang, D.H.; Hu, Y. Simultaneous sulfamethoxazole biodegradation and nitrogen conversion by *Achromobacter* sp. JL9 using with different carbon and nitrogen sources. *Bioresour. Technol.* **2019**, *293*, 122061. [CrossRef] [PubMed]
45. Nguyen, P.Y.; Silva, A.F.; Reis, A.C.; Nunes, O.C.; Rodrigues, A.M.; Rodrigues, J.E.; Cardoso, V.V.; Benoliel, M.J.; Reis, M.A.; Oehmen, A.; et al. Bioaugmentation of membrane bioreactor with *Achromobacter denitrificans* strain PR1 for enhanced sulfamethoxazole removal in wastewater. *Sci. Total Environ.* **2019**, *648*, 44–55. [CrossRef]
46. Wang, S.; Hu, Y.; Wang, J. Biodegradation of typical pharmaceutical compounds by a novel strain *Acinetobacter* sp. *J. Environ. Manag.* **2018**, *217*, 240–246. [CrossRef] [PubMed]

MDPI
St. Alban-Anlage 66
4052 Basel
Switzerland
Tel. +41 61 683 77 34
Fax +41 61 302 89 18
www.mdpi.com

Water Editorial Office
E-mail: water@mdpi.com
www.mdpi.com/journal/water

www.ingramcontent.com/pod-product-compliance
Lightning Source LLC
LaVergne TN
LVHW070631170726
843515LV00004B/231

* 9 7 8 3 0 3 6 5 0 0 3 4 8 *